AI For Horse People

From curious beginner to confident user:
A step-by-step guide to navigating AI, technology, and digital opportunities in the horse world

Jeanette Gower

Note: anywhere "he" is used, this includes "she" and "they."

Cover image: Created in ChatGPT, and extensively edited.
Thank you to Rachel Anderson for her original idea.

Although some of the images in this book were designed as examples of using Artificial Intelligence, most were from original photographs taken by my husband Peter Gower, and my good friend Janice de Gennaro, except where other photographers are mentioned.

Acknowledgements: Thank to my advance readers who offered multiple suggestions and support. Special thanks to my on-site specialist AI person, Andy, for his contributions, and my sister Frances for editing support.

First edition published in Australia.

Contents

Introduction

Why horse people need to understand AI

There is a quiet shift happening in the horse world, and it is easy to pretend it is not.

Artificial intelligence is already here. It is sitting inside search engines, social media platforms, email programs, advertising systems, photo apps, online shopping, and increasingly, everyday business tools. Most people are already interacting with AI dozens of times a day without even noticing.

And yet the moment horse people hear the words artificial intelligence, reactions tend to fall into two camps. One group becomes wildly enthusiastic while the other group immediately shuts down and assumes AI is either dangerous or something only relevant to technology companies in California.

As usual, reality sits somewhere in the middle.

This book is not about replacing horsemanship. It is not about shortcuts. And it is definitely not about handing breeding decisions over to a machine. This is about communication, organisation, clarity, and practical tools that may genuinely help you save time, improve systems, and create better outcomes for both horses and humans.

For decades, horse people built businesses through reputation, word of mouth, experience, and community connections. That still matters enormously. In fact, trust and reputation probably matter now more than ever. But the modern horse person also operates in an increasingly digital world where buyers expect information quickly, where social media shapes perception instantly, and where communication often determines whether opportunities are gained or lost.

That creates challenges. Not because horse people lack knowledge, but because many are already stretched beyond capacity. There are horses to feed, fences to repair, clients to answer, vet appointments to organise, records to manage, transport to coordinate, advertising to write, websites to update, and many other tasks happening at once while somebody messages asking "what is your best price", five minutes after you finally sit down with a cup of coffee.

Horse businesses are demanding. Most are operating with too little time and too many responsibilities. That is where AI becomes relevant.

Not as a replacement for knowledge, but as support for it.

Good horse people still need judgement. They still need feel. They still need experience, ethics, and observation. AI cannot stand quietly beside a nervous young horse and feel tension building before it explodes. It cannot recognise the subtle difference between a horse that is frightened and one that is simply confused. It cannot replace decades of practical handling experience or emotional intelligence.

Because horses are still horses, good horsemanship still matters.

What AI can do, is help horse people organise information more efficiently, communicate more clearly, edit written material professionally, generate ideas, improve educational resources, structure systems, and save significant amounts of time on repetitive tasks.

Some of the most practical applications are incredibly ordinary. They are not futuristic science fiction scenarios. There are everyday horse tasks which AI can help with remarkably well.

The important distinction is this: AI works best when paired with knowledgeable humans. Not instead of them. That distinction matters enormously throughout this book.

Because there are also areas where AI should never replace people. Welfare decisions, veterinary emergencies, breeding ethics, horse assessment, and emotional judgement still belong firmly in human hands. Responsible horse people must understand both the strengths and the limitations of these tools.

Used thoughtfully, AI can support good systems. Used carelessly, it can create confusion, misinformation, and unrealistic marketing very quickly. That is why this book exists, to help horse people understand how these tools can be used practically, ethically, and responsibly within the real world of horses.

And importantly, to explain AI in normal language.

Because most horse people do not want corporate jargon or complicated technical explanations.

You do not need to understand coding. You do not need expensive software, complicated equipment, or a university degree in technology. What you do need is curiosity, common sense, ethical judgement, and a willingness to experiment gradually.

The horse industry has always rewarded people who adapt thoughtfully.

Throughout this guide, we will explore practical ways AI can support you without losing your authenticity or integrity. We will cover advertising, branding, organisation, educational content, stallion promotion, social media, buyer communication, and ethical concerns surrounding AI use in different ways.

Some ideas will suit you immediately. Others may not suit you at all. That is perfectly fine. The goal is not blind adoption. The goal is thoughtful understanding.

Because whether people embrace AI enthusiastically or cautiously, one thing is already becoming clear. The horse industry is entering a new era of communication. The breeders, trainers, businesses, and horse people who learn how to combine traditional knowledge with modern tools will likely be the ones best positioned for the future.

Artificial intelligence is already here

***

Multitasking is not optional in horse keeping: it is the job description.

Chapter 1

What AI actually is

Software designed to process information

Artificial intelligence sounds far more intimidating than in fact it is. The name alone tends to make people imagine robots taking over the world, computers becoming emotionally unstable, or machines suddenly deciding they know more about horses than breeders who have spent forty years in the industry.

Thankfully, we are not there.

In simple terms, AI is software designed to process information, recognise patterns, and generate useful responses based on the information and instructions it receives. The easiest way to think about AI is this: it is a very fast assistant.

It reads quickly, organises quickly, summarises quickly, writes quickly, and responds quickly. But it still depends entirely on the quality of information and direction it receives from humans.

It is a tool that predicts language and information patterns based on the instructions it is given. Sometimes it does that brilliantly. Sometimes it produces complete nonsense with the confidence of somebody explaining horse nutrition after reading half a Facebook comments section.

This is why learning how to use AI properly matters. The quality of the outcome depends heavily on the quality of the question.

AI versus search engines

This is one of the easiest ways to understand the difference. Google helps you find information. AI helps you work with information. That distinction changes everything.

For years, people searched online by typing questions into search engines and sorting through websites manually. A breeder might Google how to write a horse sales advertisement, then spend half an hour opening articles, reading random advice, becoming distracted by saddle pads, and somehow ending up researching fencing by accident.

Horse people are especially talented at accidental research detours, what they call the 'rabbit hole.'

AI works differently. AI responds directly to your request (or prompt):

> *Help me write a sales advertisement for a rising 4-year-old Morgan mare. Quiet to handle, green under saddle, best suited to experienced home wanting a future all-rounder. Avoid hype or exaggerated language.*

Within seconds, you receive a draft. Not perfect. Not final. But useful.

You then adjust it using your own knowledge and judgement.

AI is already everywhere

Many people believe AI is some futuristic technology that has suddenly appeared out of nowhere. In reality, most people have already been using AI for years without thinking about it.

Examples include predictive text on phones, autocorrect, suggested replies in emails, social media algorithms, voice assistants, online shopping recommendations, GPS navigation systems, photo editing tools, and spam filters.

The difference now is that AI tools have become conversational and interactive. Instead of simply correcting spelling quietly in the background, AI can now help you brainstorm ideas, structure documents, edit writing, organise information, summarise articles, generate educational content, and simplify repetitive tasks.

That shift is what makes AI particularly practical for horse people. The difference between GPS and AI programmes is the difference between a tool that works quietly in the background and one you can actually have a conversation with. Passive AI made decisions for you, invisibly. Interactive AI lets you direct it, ask questions, give instructions, refine results, and shape the outcome.

That is an entirely new kind of tool, and it is why so many people are only now beginning to realise what is possible.

What AI does well

AI performs well when handling communication, organisation, brainstorming, and simplification.

It can write drafts, edit wording, improve clarity, generate headings, summarise information, sort ideas, create systems, format information, structure records, generate checklists, explore alternatives, and explain complex ideas in plain language. These are practical applications, not science fiction.

What AI does badly

AI also has weaknesses, and responsible horse people need to understand those clearly. AI struggles with real world judgement. It cannot physically assess horses. It cannot evaluate movement properly from lived experience. It cannot feel tension, softness, anxiety, or responsiveness in the way experienced handlers can about their horses.

AI also cannot replace emotional intelligence. It cannot truly understand human emotions or horse behaviour dynamics with the depth of lived experience. It can imitate language about those things, but imitation is not the same as wisdom.

AI can also be inaccurate. Sometimes it produces incorrect information very confidently. This is often called *hallucination*, which sounds far more magical than it really is. It means that AI generated inaccurate information while sounding extremely certain about it. Rather like that one person at every horse event confidently giving terrible advice near the coffee cart.

In practice, this can look like an AI confidently naming a specific Welsh Cob sire as the foundation of a particular bloodline, when that sire never existed. Or describing a breed characteristic with complete authority that is absolutely wrong. The information sounds plausible, is written clearly, and is entirely invented. That is why you should never use AI as the final word on bloodlines, health information, breed standards, or anything factual where accuracy genuinely matters.

AI should never replace ethics and welfare decisions. Breeding choices, veterinary emergencies, horse welfare, and safety concerns remain human responsibilities.

Why prompts matter so much

One of the biggest frustrations beginners experience with AI is getting weak or generic responses. Usually this happens because the prompt was too vague.

For example: *Write horse ad.* That gives AI almost nothing useful to work with. Now compare that to:

> *Help me write a sales advertisement for a quiet 14.2hh family gelding suitable for confident teenagers and adult riders. Include temperament, manners, and current training level. Avoid hype and unrealistic claims.*

12 myths horse people can safely ignore

Whenever a new technology appears, myths quickly follow. Artificial intelligence is no different. Before exploring what AI can do, it helps to clear away some of the most common misconceptions.

Myth 1: AI Knows Everything
AI can access enormous amounts of information, but it does not know everything. It can make mistakes, miss context, and occasionally provide incorrect answers. Treat it as a helpful assistant, not an unquestionable authority.

Myth 2: AI Will Replace Horse People
AI can help with administration, communication, and organisation, but it cannot replace horsemanship, judgement, experience, or relationships.

Horses still need knowledgeable people.

Myth 3: Only Young People Can Learn AI
Many of the people who benefit most from AI have decades of experience with horses. Technology skills can be learned. Experience takes years to acquire and remains incredibly valuable.

Myth 4: You Must Be Technical
Most AI tools work through ordinary conversation. You do not need to become a programmer. If you can ask questions, you can begin using AI.

Myth 5: AI Is Either Perfect or Useless
Like any tool, AI has strengths and limitations. It can be excellent for brainstorming, writing, and organisation, while still requiring human judgement and verification.

Myth 6: You Need Every New Tool
You do not need dozens of apps and subscriptions. Most people achieve excellent results with a small number of tools used consistently.

Myth 7: Traditional Knowledge No Longer Counts
As information becomes easier to access, practical experience becomes even more important. Good judgement and horse knowledge remain invaluable.

Myth 8: AI Is a Shortcut to Expertise
AI can support learning, but it cannot replace experience, observation, practice, and reflection. Expertise is still earned.

Myth 9: AI Is Only for Businesses
Private horse owners can benefit from AI too. It can help with learning, planning, organisation, record keeping, and research.

Myth 10: If You Haven't Started Yet, You're Behind
There is no race. Many people are only beginning to explore AI. Start where you are, learn at your own pace, and focus on what is useful to you.

Myth 11: AI is programmed to give back only what you want to hear.
AI is designed to be accurate and helpful, not flattering. It will contradict you, correct your assumptions, and tell you when your prompt is unclear, which is not the behaviour of a system built to flatter.

Myth 12: AI is programmed to continue engagement with you, which rewards the companies behind it.
While AI companies do benefit from engagement, the tools themselves are not designed to manufacture dependence. A good AI session ends when your task is done.

AI is not replacing horse people

This fear comes up constantly, and it is understandable, but good horsemanship is still deeply human. A horse still needs patience, consistency, observation, timing, 'feel', and emotional regulation. No machine replaces that.

What AI does is remove some of the repetitive communication and administrative pressure surrounding horse businesses so people can focus more energy on the horses themselves. That is a very different thing.

Start simple. Do not over-complicate this.

You simply need to begin experimenting with small practical tasks. Ask AI to improve wording in a sales ad, edit grammar in a brochure, generate headings for a website, summarise a veterinary article, create a feeding checklist, rewrite an awkward email professionally, or brainstorm social media ideas.

That is enough. Start with one task. Do it imperfectly. Adjust. Try again. That is the whole method.

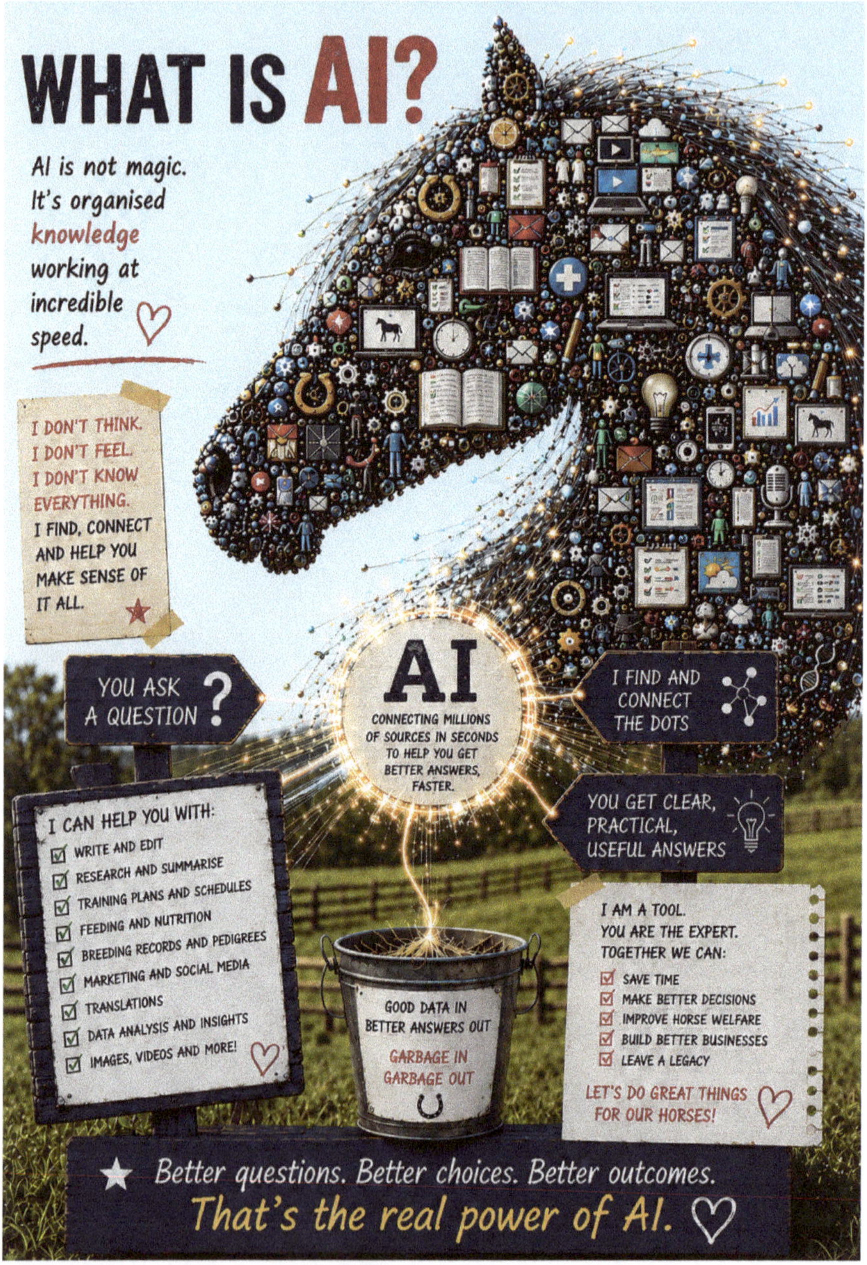

AI is not magic and it is not a threat. It is organised knowledge working at speed, and in the right hands, it is a powerful tool for anyone in the horse world.

Chapter 2

Where AI helps horse people most

The surprisingly ordinary

The most effective use of AI is often surprisingly ordinary. It is not about turning your stable into a technology company. It is about reducing friction in the everyday tasks that quietly consume enormous amounts of time and mental energy.

Horse people are already managing an extraordinary number of responsibilities: feeding schedules, breeding records, registrations, advertising, transport coordination, social media, websites, invoices, competition entries, vet reports, educational material, emails, and paperwork. Most people are holding entire systems together through memory, habit, screenshots, notebooks, and what can only be described as organised chaos.

Many horse businesses survive because horse people are incredibly resourceful under pressure.

But survival mode is exhausting. This is where AI becomes genuinely useful.

AI helps most where communication matters

Communication is probably the single biggest area where AI can improve horse businesses immediately. Most horse people are excellent at horse-related knowledge but often struggle to communicate that knowledge clearly, consistently, and professionally online.

Because writing is difficult, especially when you are tired, busy, emotionally invested in your horses, and trying to answer messages while carrying feed buckets.

Writing sales advertisements

This is one of the easiest and most valuable places to begin. Most horse advertisements fall into one of three categories. Too vague, too emotional, or wildly exaggerated.

Instead of staring blankly at a screen wondering how to describe a horse professionally, you can provide real details and ask it to structure the wording calmly and honestly.

Help me write a credible sales advertisement for a 13.hh 3 year old quality black Riding Pony gelding. Quiet on the ground, green under saddle, best suited to an experienced rider wanting a future competition horse. Avoid hype or unrealistic claims.

Immediately, the pressure reduces. You are no longer starting from nothing. And honestly, that alone saves many people enormous mental energy.

Editing brochures and written material

Many owners and horse businesses already have relevant information written down somewhere. The problem is that much of it sounds rushed, repetitive, unclear, or inconsistent because it was written quickly between so many other responsibilities.

AI is excellent for improving clarity without changing the core message.

Please improve the wording of this brochure while keeping it warm, honest, and professional. Do not exaggerate the horses or sound overly sales focused.

You then paste your original wording underneath. This works beautifully for breeding brochures, stallion flyers, website information, educational handouts, buyer information packs, client welcome documents, and newsletters.

Rewriting awkward emails professionally

Horse industry communication can become emotionally exhausting, especially if you're dealing with difficult buyers, unrealistic expectations, or messages that arrive with all the warmth of a legal threat.

Please rewrite this email in a friendly and polite manner while keeping firm boundaries.

Help me respond professionally to a buyer enquiry that is not suitable for this horse.

That pause between emotional reaction and professional response can protect reputations enormously.

Creating headings and taglines

Many people freeze completely when trying to create headings. Whether it is sales advertisements, Facebook posts, stallion promotions, website pages, or educational content, AI is very effective at brainstorming options quickly.

Give me 15 calm and business-like heading ideas for a family horse breeding programme focused on temperament and trainability.

Suggest Facebook advert headings for a Welsh Cob stallion standing at stud.

You may only need one suggestion, but sometimes one good idea is all you need to regain momentum.

AI is excellent for organisation

Horse people carry extraordinary amounts of information mentally. Most of this knowledge exists inside human memory rather than organised systems.

I will paste notes about several mares. Please summarise recurring patterns, strengths, and management observations.

That does not replace breeder judgement. But it can help identify patterns more clearly and organise information into usable formats.

This becomes especially helpful for broodmare records, feeding schedules, foal development notes, veterinary tracking, buyer communication, stable management systems, and competition records.

AI can simplify educational content

Breeders, trainers, and instructors will explain the same concepts repeatedly to clients and buyers. Feeding routines, transition advice, training expectations, management requirements, and suitability discussions all need careful communication. AI can help turn those explanations into reusable resources.

Help me create a helpful transition guide for buyers taking a young horse to a new home.

Explain EMS in horses in simple terms for first time owners.

Horses benefit when owners receive clear, calm, realistic information from the beginning.

AI helps with social media content too

This does not mean turning your horse page into robotic motivational nonsense filled with glitter graphics and inspirational quotes floating over sunsets. AI can help generate educational content ideas and improve consistency online.

> *Give me 10 educational Facebook post ideas for a horse breeding programme focused on temperament and responsible management.*
>
> *Help me write a calm and informative Facebook post explaining why suitable rider matching matters.*

This helps businesses stay visible and informative without constantly struggling for ideas.

AI can save time when used thoughtfully

Artificial Intelligence is not useful when people expect it to magically run their business without input, judgement, or oversight. That usually creates generic nonsense very quickly. The people who benefit most are generally already thoughtful, ethical, and organised, because AI amplifies existing habits.

Start with the tasks you already dislike. Do not start with complicated projects. Start with the repetitive tasks that drain energy unnecessarily. Maybe that is writing Facebook captions, editing grammar, creating headings, organising notes, drafting brochures, rewriting emails, or structuring handouts.

Small wins matter. Confidence builds through use. Exactly the same way confidence develops in horses. Quiet repetition. Consistent exposure. Low pressure learning. People are not nearly as different from horses as they like to believe.

The real value of AI for horse people

At its best, AI gives horse people something incredibly valuable: mental breathing room.

Less staring at blank screens. Less rewriting awkward sentences repeatedly. Less procrastination over communication tasks. Less time lost trying to organise scattered information.

Where AI should not replace humans

One of the biggest problems surrounding AI right now is that people tend to swing wildly between two extremes. Some believe AI will solve everything. Others believe it will destroy civilisation by Thursday afternoon. Neither position is especially helpful.

The reality is far more practical. AI is merely a tool. A hammer is a tool. You can use it to build projects carefully, or you can smash your thumb. Similarly, AI is a very good tool in some areas, and a very dangerous tool in others.

A horse is not a spreadsheet. Good horsemanship is not only knowledge. It is emotional intelligence, observation, body language, timing, patience, and practical experience combined. That remains entirely human. Despite all the rapid technological changes happening around us, horses still require deeply human skills.

AI should never make breeding decisions

AI can help organise breeding records. It can summarise observations. It can assist with communication and documentation. But it should never replace breeder judgement.

Good breeding decisions involve far more than data. Many of the best breeders in the world notice things that are difficult to even explain verbally; tiny behavioural patterns, subtle consistency between bloodlines, the difference between a horse that merely performs well and one that genuinely passes on quality minds, generation after generation.

AI is not a veterinarian

This should be obvious. Unfortunately, the internet has already proven repeatedly that obvious things sometimes require saying loudly. AI can help explain veterinary concepts in simpler language. It can summarise articles. It can clarify terminology and help owners understand information more clearly. But it should never replace proper veterinary care.

If your mare is colicking at midnight, this is not the moment to ask a chatbot philosophical questions about gut health. Call the vet.

The same applies to injuries, emergencies, medication decisions, breeding complications, and welfare concerns. AI can support education. It cannot physically assess horses. That distinction matters enormously.

AI cannot replace emotional intelligence

Observers constantly assess people as much as horses. They notice buyer suitability, emotional maturity, confidence levels, handling ability, unrealistic expectations, risk factors, and stress responses.

A horse may technically suit somebody on paper while clearly being the wrong emotional fit in reality. AI cannot fully interpret those subtleties. It does not truly understand emotional complexity, body language, intuition, or lived social experience the way humans do.

Confident decisions based on instinct developed through decades of observation are still what makes the best horsemen.

AI generated horse images are becoming a real problem

AI generated horse photos are already appearing online, and many people cannot immediately recognise them as fake. That creates enormous ethical concerns. Some images show impossible conformation, impossible movement, impossible lighting, and impossible perfection.

While they may look visually impressive at first glance, they distort reality. Horse advertising already struggles with unrealistic presentation. AI generated imagery risks making that problem significantly worse.

That is why this book strongly advocates using real horses and real photographs. Honest presentation protects buyers, protects reputations, and most importantly, protects horses.

AI can create confidence people have not earned

This is another subtle danger. AI can help people sound knowledgeable very quickly. Sometimes too quickly. A person with limited experience can suddenly produce polished educational material, professional- sounding advertising, and confident responses that appear far more informed than their actual horse knowledge.

Polished wording does not automatically equal expertise. Polished branding does not automatically equal good horsemanship.

Critical thinking still matters. Research still matters. Reputation still matters. Remain sceptical of anyone who appears online overnight claiming expertise with suspiciously perfect marketing.

Can you teach an old dog new tricks?

The horse industry has always attracted confident storytellers. AI simply gives them better tools. But this book is here to tell you that anyone can learn to use AI. The strongest use of AI is support.

The people who will likely benefit most from AI are not shortcut seekers. They are thoughtful horse people already trying to improve systems, communication, and consistency responsibly. Because tools amplify existing habits, ethical people tend to use tools ethically. Disorganised people often create faster chaos.

A horse float can transport horses safely or dangerously. A whip can guide softly or create fear. Technology is no different.

What does it cost?

One of the first questions you might ask is whether artificial intelligence is expensive. The good news is that most AI tools offer free versions, and for most beginners those free versions are more than adequate. They allow you to experiment, learn, and discover what is incredibly helpful without spending any money.

As your confidence grows, you may decide to upgrade to a paid version if additional features become valuable. However, there is no need to begin there. The goal is to learn first and spend later, if you decide the benefits justify the cost.

Where do you sit?

The following checklist helps you see clearly where you are right now, and where the most useful starting points might be. There are no wrong answers.

A Horse Person's AI Readiness Checklist

One of the reasons artificial intelligence feels overwhelming is that people often assume they need to do everything at once. This checklist is designed to help you assess where you are today and identify practical next steps.

Can you answer yes to the following?

☐ I am willing to learn new skills.
☐ I am curious about emerging technology.
☐ I understand that AI is a tool, not magic.
☐ I am comfortable experimenting without expecting perfection.
☐ I understand that mistakes are part of learning.
☐ I recognise that technology will continue changing.
☐ I believe experience and judgement still matter.

Basic Skills

☐ I can use email confidently.
☐ I can search for information online.
☐ I can save and locate documents.
☐ I can upload photographs or files.
☐ I can navigate websites comfortably.
☐ I can use a smartphone, tablet, or computer independently.

Artificial Intelligence

☐ I have experimented with an AI chatbot.
☐ I have asked AI questions about subjects that interest me.
☐ I understand that AI sometimes makes mistakes.
☐ I verify important information.
☐ I know how to improve a prompt if the first answer is unhelpful.
☐ I feel comfortable having conversations with AI.

Horse Business Systems

☐ I have organised records.
☐ I have client information stored appropriately.
☐ I use calendars or scheduling systems.
☐ I have written policies or procedures.
☐ I have suitable templates for repetitive tasks.
☐ I can easily find important information when needed.

Communication

☐ I have a clear way of communicating with clients.
☐ I answer common questions consistently.
☐ I have written information available for clients.

☐ I use email effectively.
☐ I understand the value of newsletters.
☐ I communicate expectations clearly.

Digital Presence

☐ I have a website or online presence.
☐ My contact information is easy to find.
☐ My services are clearly explained.
☐ My photographs represent my activities authentically.
☐ My online information is current.
☐ I understand how people discover me online.

Knowledge Preservation

☐ I maintain records.
☐ I document useful information.
☐ I preserve photographs and documents.
☐ I organise educational resources.
☐ I consider what can be preserved for future generations.
☐ I recognise the value of my own experience.

Educational Opportunities

☐ I regularly learn new things.
☐ I enjoy improving my knowledge.
☐ I read educational material.
☐ I attend clinics, workshops, or events.
☐ I share knowledge with others.
☐ I remain curious. Curiosity remains one of the strongest indicators of future success.

Practical Next Step: Choose one item from the checklist that you could improve upon during the next thirty days.

What your checklist results mean?

Not everyone starts in the same place. Your checklist results point you toward the chapters that you need first. These three people completed the same checklist. See which one is most like you.

Profile 1: Sarah — The Riding Instructor

Mindset ✓ | Systems ✓ | Digital Presence ✗ | AI Knowledge ✗

Sarah has been teaching lessons for eighteen years. Her horses are well cared for, her students are loyal, and she knows what she is doing. But her booking system is a whiteboard, her lesson notes live in her head, and she has never posted consistently on Facebook. The idea of using AI feels overwhelming.

Her next step: Start with the lesson template chapter.

Why this order makes sense: Sarah does not need to understand AI before she benefits from it. She needs one small win, a ready-made lesson plan she can adapt in five minutes instead of starting from scratch.

Profile 2: James — The Breeder

Mindset ✓ | Systems ✓ | Digital Presence ✗ | AI Knowledge ✗

James runs a respected breeding operation. His records are meticulous, his pedigree knowledge is deep, and he has a waiting list for foals. He has never needed social media because word of mouth was always enough, until now. The market has shifted, buyers are younger, and sales are going to breeders with half his experience but twice his online presence.

His next step: Go straight to the sales ads and promotional chapters.

Why this order makes sense: James already believes in the value of what he offers and already has solid systems. What he needs is to translate his expertise into words that work online. The AI tools for writing sales descriptions, mare and foal announcements, and stallion promotions are exactly where he will see the fastest return.

Profile 3: Maria — The Business Owner

Mindset ✓ | Systems ✓ | Digital Presence ✓ | AI Knowledge ✓ (mostly)

Maria runs an equestrian centre with trail rides, lessons, agistment, and a small retail corner. She is already using a booking platform, posts regularly on social media, and has experimented with AI tools. She is not overwhelmed, she is under-challenged. She knows she is not getting everything out of the tools she has.

Her next step: Go straight to the advanced chapters on video content, efficiency systems, and using AI from your phone.

Maria already knows what she is doing. What she needs are ideas she has not considered yet. The chapters on short-form video, batching content, and ethical AI use in reputation management are where she will find genuine value.

The pattern

This checklist is not measuring how good you are at your job. Sarah, James, and Maria are all excellent at what they do. What it measures is where the gaps are, and gaps are simply the next place to grow.

Gaps across the board? Start with mindset, then systems, then presence, then AI tools. Each layer supports the next.

There is no wrong result. There is only where you are right now, and where you are headed. Build your horse world with the same patience and intention you bring to your horses, and it will carry you further than you expect.

Chapter 3

Setting up your first AI tools

You can skip this chapter if you have already started using AI

The internet is filled with articles, videos, opinions, reviews, rankings, and endless discussions about which tool is best. That can make a simple process seem far more complicated than it really is.

Step 1: Choose one tool

For most horse people, ChatGPT is usually the easiest place to begin. Once you become comfortable, you can explore other tools if they offer something useful.

Step 2: Create an account

To access it, open your internet browser and visit the ChatGPT website. You will be given the option to create an account or sign in if you already have one. The registration process is like many other online services.

https://chatgpt.com

Typically, you will enter your email address, create a password or use an existing Google, Apple, or Microsoft account. Verify your email address if requested, then sign in.

Once you are in, you will see a text box. That is it.

It works exactly the way a chat window does. You type something, the AI responds. There is nothing technical to configure, no dashboard to navigate, and nothing that requires any special knowledge. If you have ever sent a message on Facebook or replied to an email, you already know how to use it.

If ChatGPT is unavailable in your region or you prefer another platform, Claude, Gemini, and Perplexity all work the same way. The interface may look slightly different, but the experience is identical: you type, it responds. Pick whichever one feels comfortable and stick with it.

Free versus paid accounts

Most platforms offer both. My recommendation would be to start free. Learn first. Upgrade later if a tool consistently saves time or provides value. You will discover that free accounts are sufficient for quite some time. The goal is merely to get started.

Step 3: Complete your profile by customising your settings

The idea is to create your 'voice' to sound like you, a conversation, not a lecturer, reporter, or a research paper. You may be asked your name, preferred language, work interests, communication preferences.

You do not need to complete every option immediately. You can revise this as you go along to reflect your voice more accurately.

Step 4: Start a Conversation

Treat AI as though you were speaking with an assistant. Don't overthink it. Learn through real problems. The fastest learning happens when you solve actual problems. The more relevant the task, the faster your confidence grows. Try:

Help me create a riding school newsletter

What information should be included in an agistment welcome pack?

Help me develop a pedigree template over 5 generations

Explain why you should never breed two overo horses together for a short article on my website

Privacy

When you type something into an AI chat box, that text is sent to the company's servers and used to generate your response. On free accounts, it may also be used to improve the AI's training, which means information you paste could, in theory, become part of what the system learns from. Most reputable platforms have privacy policies explaining this, but the practical takeaway is simple: treat a free AI chat window like a public space, not a locked filing cabinet.

Avoid uploading:

- confidential client information
- financial records

- legal documents
- sensitive personal information

Paid accounts generally offer stronger privacy protections. Many explicitly state that your conversations will not be used for training, and some offer enterprise-level confidentiality agreements. If you are dealing with client records, veterinary details, breeding contracts, or any information that belongs to another person, either use a paid plan with clear privacy guarantees or leave names and identifying details out entirely.

Bundled platform caution

As AI becomes more popular, you will increasingly come across "all-in-one" platforms promising to combine several AI tools into a single subscription. These often bundle text generation, image creation, video tools, and social media scheduling together, sometimes marketed with names you have never heard of, and sold through one-off "lifetime deal" pricing that sounds appealing because it avoids ongoing monthly fees.

It is worth understanding what these platforms actually are. In most cases, they are not building their own AI. They are wrapping access to the same underlying models, Claude, ChatGPT, or similar, inside their own interface, then charging you for that interface. Sometimes this genuinely saves time by bringing several tools into one place. But the underlying AI is usually something you could access directly, often more reliably and at lower cost.

The bigger concern is longevity. Running AI tools costs the provider money every time someone uses them, regardless of how the platform was sold to you. A "lifetime" price covers none of those ongoing costs in any sustainable way, which means some of these platforms quietly reduce quality, restrict usage, or disappear altogether once the initial sales revenue is spent. If you have invested time building content, client systems, or branding inside one of these platforms, losing access can be far more disruptive than losing a simple chat history.

None of this means bundled platforms are automatically dishonest. But before committing money, particularly to lifetime pricing, ask yourself what this platform does that you cannot already do directly, who is behind it, and what happens to your work if the platform closes down in two years. As with everything else in this book, the goal is not to avoid these tools out of fear, but to look at them with the same common sense you would apply to any other purchase for your horse business.

> *Act as a researcher helping me evaluate an AI platform or bundled subscription service before I commit money to it. The platform is called [insert name], and here is what it claims to offer: [paste description or website text].*
>
> *Please research and tell me: what company is actually behind this platform, where are they based, and how long have they been operating. Are there any*

red flags such as unclear ownership, recent name changes, or complaints about cancelled subscriptions or lost access.

What underlying AI models does the platform use, and is it building its own technology or repackaging existing tools.

What do independent reviews say, separate from reviews on sites that may earn referral fees for recommending it.

Finally, based on what this platform does, is there a simpler or more direct way I could achieve the same result without it.

Build a prompt library

As you begin experimenting, you will occasionally receive responses that are particularly useful. Save the prompt. Create a folder. Start a document. Experienced users rely heavily on collections they have developed over months or years.

It is one of the smartest habits you can develop. A prompt library is a collection of instructions that consistently produce useful results.

The beauty of a prompt library is that it grows over time. Each successful prompt becomes an asset that can be reused whenever needed. In this book I have provided many that you can use directly into AI.

Whenever you discover a prompt that works well for you, save it. Over time you build your own personalised AI toolkit.

Act as my AI on-boarding assistant. I have just created a new AI account and would like help configuring it for horse related projects.

Please walk me through the setup process one step at a time. Help me understand profile settings, choose useful preferences, create my first conversation, save relevant prompts, organise future projects, and avoid common beginner mistakes.

Explain everything in plain English and assume I have never used AI before.

Chapter 4

Choosing yourAI toolbox

Different tools for different jobs

Understanding what the tools do can help you recognise opportunities when they arise. Some are better suited to particular tasks than others.

ChatGPT

For beginners, ChatGPT is likely to become the first and most frequently used AI tool. Its greatest strength is versatility. It can help write newsletters, create educational resources, draft policies, brainstorm ideas, organise information, build websites, prepare horse sale advertisements, create welcome packs, and assist with planning.

It is often the equivalent of having a patient assistant available whenever you need one. I frequently recommend ChatGPT as a starting point because it does so many things well. You may eventually discover specialist tools that perform certain tasks better. That is perfectly normal. But ChatGPT remains one of the best places to begin.

Best suited for: writing, planning, communication, newsletters, educational resources, brainstorming, websites, general problem solving.

Cost: Free and paid plans available. **Link:** http://chatgpt.com

> *Help me create a suitable welcome pack for new agistment clients. Include services, expectations, emergency procedures, payment information, and frequently asked questions.*

Claude

If ChatGPT is a general assistant, Claude often feels like an editor. Claude is particularly good at handling large amounts of text because it is so much faster at it than ChatGPT. I use Claude a lot for searching through long material and editing. Claude doesn't handle image creation well. It can give you the prompt, but then I would put the prompt into ChatGPT or Midjourney.

Claude has recently introduced Claude Design, which handles image *editing* well.

> *From this PDF, select headings which should be changed from title case to sentence case.*

This makes it useful for writers, breed societies, organisations, policy development, long reports, educational material, and manuscript projects. Claude is especially helpful when refining existing material rather than creating it from scratch. If you are writing a breed history, preparing a handbook, creating educational resources, or editing a book, Claude is worth exploring.

Best suited for: editing, long documents, reports, policies, handbooks, manuscripts, educational resources.

Cost: Free and paid plans available. **Link:** http://claude.ai

> *Please review this breed history article and improve clarity, structure, readability, and flow while preserving the original meaning.*

Perplexity

Perplexity is best thought of as a research assistant. While ChatGPT and Claude often generate answers from what they already know, Perplexity is particularly resourceful when you want current information and sources.

It can help locate information, summarise articles, compare viewpoints, and provide references for further investigation, useful when researching horse industry topics, regulations, organisations, events, historical information, or emerging trends.

It is important to remember that research still requires judgement. Perplexity helps you find information. It does not replace critical thinking.

Best suited for: research, current information, summaries, source discovery, background information.

Cost: Free and paid plans available. **Link:** http://perplexity.ai

> *Please provide an overview of endurance riding in [my country], including major organisations, recent developments, and relevant resources for further reading.*

Gemini

Gemini is Google's AI assistant. Its greatest advantage is integration with Google's ecosystem. If you already use gmail, Google docs, Google drive, Google sheets, or Google calendar, Gemini may feel familiar.

Many people appreciate the convenience of having AI integrated into systems they already use.

Best suited for: Google users, planning, writing, productivity, research.

Cost: Free and paid plans available. **Link:** http://gemini.google.com

Help me organise a six-month content plan for our local breed society newsletter and educational program.

Microsoft Copilot

Copilot occupies a similar position within Microsoft's ecosystem. If you regularly work in Word, Excel, Outlook, or Powerpoint, Copilot, can provide assistance directly within those programs. For organisations already using Microsoft products, this can be very convenient.

Best suited for: Word documents, spreadsheets, email, presentations, business administration

Cost: Usually linked to Microsoft subscriptions. **Link:** http://copilot.microsoft.com

Help me organise this horse event budget spreadsheet and identify areas that may require additional planning.

Midjourney

Midjourney is different from the other tools discussed so far. Its specialty is image generation. Rather than producing text, Midjourney creates images based on written instructions. This can be excellent for concept illustrations, marketing ideas, promotional graphics, educational material, and creative projects.

Best suited for: image creation, concepts, illustrations, visual ideas.

Cost: Paid subscription. **Link:** http://midjourney.com

Create a photo-realistic grey Arabian horse standing in soft morning light beside a rural fence line, highly detailed, realistic photography style.

Keep it simple. Everything else can wait until you discover a genuine need.

Advanced prompt:

Theme: AI tools

Suitable for: beginners wanting to get started

Act as an AI technology advisor for the horse industry. Please recommend the most suitable AI tools for the following goals: [Insert goals]

Include: recommended tools, learning difficulty, approximate cost, advantages, limitations, beginner recommendations.

Please focus on practical solutions rather than complicated technology.

Be careful what you wish for

AI images have improved remarkably since I tried this out about 2 years ago, but they sometimes still contain anatomical inaccuracies, unusual tack, extra legs, missing legs, or other oddities that become obvious once you know what to look for.

Horse and human bond. I asked AI to remove the tack. This is what I got! That was only 2 years ago, so much has changed.

Chapter 5

Prompting AI

How to ask better questions

One of the biggest reasons people become frustrated with AI is surprisingly simple. They ask terrible questions.

Not because they are unintelligent, but because most people are still learning how AI actually works. Horse people are not alone there. Businesses across every industry are fast discovering that the quality of AI responses depends heavily on the quality of the instructions being given.

This is called *prompting*.

Think of it like giving instructions to a new groom who is keen to help but knows nothing about your property, your horses, your routines, or your standards. The more clearly and specifically you explain what you need, the more useful they become. Vague instructions produce vague results, not because they are incompetent, but because they have nothing solid to work from. AI is exactly the same.

A prompt is the instruction or question you give the AI. Learning how to write better prompts changes everything because AI is not a mind reader. It does not automatically know what type of horse you are selling, what tone you want, whether you value realism over showmanship, who your audience is, what details matter most, or what style of communication suits your business.

You need to tell it clearly.

Why vague prompts create weak results

Most beginners start with prompts like this: *Write a horse ad for my Welsh Cob gelding.*

Technically, AI can respond to that. But the result will almost always feel generic, awkward, or unrealistic because the AI has been given almost no context.

It is a bit like asking somebody to recommend a good horse without explaining whether you need a child's pony, a campdrafter, a dressage horse, or something safe enough for your nervous aunt who still believes every horse over 14hh is scary. Specificity matters.

Now compare that vague prompt with this:

> *Help me write a calm and honest sales advertisement for a rising 6-year-old 14.3h Welsh Cob gelding. Quiet to handle, well-trained foundations, demonstrating versatility and scope over jumps. Best suited to an experienced home wanting a teenager's all-rounder. Avoid hype or exaggerated language. Ask me questions to complete the advert.*

Completely different outcome.

The AI now understands the horse type, age and training stage, temperament, intended audience, desired tone, and ethical boundaries. It will ask you further questions such as if you want to include price or location.

The clearer your communication becomes, the stronger the output usually becomes. If you give vague instructions, you will receive vague work. If you give detailed instructions, examples, context, and clear expectations, the work improves significantly.

Getting the best out of AI: four things that matter

Every good prompt should include:

Task: What do you want AI to do? Write, rewrite, summarise, brainstorm, edit, simplify, or organise? Be specific about the action.

Context: What horse? What business? What audience? What is the purpose of this piece of communication? The more detail you give here, the more useful the result.

Tone: How do you want it to sound? Professional, warm, excited, educational, honest, friendly, luxury focused, or family oriented? AI defaults to a generic voice unless you tell it otherwise.

Boundaries: This is where ethical communication becomes important. You can ask AI to avoid hype, avoid exaggeration, keep wording realistic, use plain language, or avoid sounding overly sales focused. These instructions protect your integrity and keep the output consistent with your values.

Practical examples you can use immediately

> Weak: *Write a Facebook post about temperament in horse breeding.*

Better: *Write a calm and informative Facebook post explaining why temperament matters in horse breeding. Tone should feel warm, educational, and practical rather than overly emotional. Make it short and reflective of my experiences.*

See the difference immediately? The second version gives direction. Direction creates better outcomes.

Brochure editing

You may already have good written material but struggle with presentation. Instead of rewriting everything manually, try this:

Please improve the wording of this breeding brochure attached, while keeping it warm, honest, and factual. Do not exaggerate the horses or make unrealistic promises.

Then paste or upload your original wording. The programme will provide a plus sign or drop-down menu to attach the file.

This works beautifully for brochures, websites, information packs, buyer education sheets, stallion flyers, and social media posts. The AI is not replacing your knowledge. It is helping present your knowledge more clearly.

Headings and taglines

AI saves enormous time with headings because you may hit a blank when trying to create them.

Give me 15 calm and business-like headings for a horse breeding programme focused on trainability, temperament, and family horses.

Suggest taglines for a Welsh Cob stud focused on traditional bloodlines and sensible horses.

You may only need one idea. But one excellent idea can save thirty minutes of frustration.

AI responds well to examples

You can show AI examples of the tone or style you like. The more guidance you provide, the more refined the response becomes.

Rewrite this advertisement in a friendly, mentoring style like an article rather than aggressive sales language. Make this wording feel more educational and trustworthy rather than pushy. Keep it factual, no more than 3-4 paragraphs.

Ask AI to simplify things

This is one of the most practical uses for horse people, especially when dealing with complicated veterinary information, contracts, or educational material.

Explain this veterinary article in plain language suitable for first time horse owners.

Please re-frame this breeding contract into simple, every-day language.

This becomes incredibly helpful for buyer communication and educational resources.

Refine responses repeatedly

You can continue refining the response. This process often produces the strongest results. And this is where you may suddenly realise AI becomes genuinely useful rather than intimidating.

Make this wording warmer.

Shorten this version.

Add more information about temperament.

Remove repetitive wording.

Make this sound more polished but still friendly.

Start simple and practical

Do not over-complicate this process. You do not need perfect prompts. You only need to begin. Rewrite an awkward email. Improve a sales advertisement. Create brochure headings. Generate social media ideas. Simplify educational information. Organise stable notes.

The more you use AI, the more naturally prompting starts to develop. Eventually, you stop feeling like you are using technology and start feeling like you have a valuable personal assistant.

Act as an expert prompt engineer and horse industry communications consultant. Please improve the following prompt so that it produces more accurate, useful, detailed, and informative results. Identify missing information, unclear instructions, assumptions, and opportunities for improvement.

Then create a beginner version, an advanced version, and a horse business version. [insert prompt text]

Where you see brackets such as [insert here], this means you need to insert something into the prompt. For example, I might generate a prompt relating to a particular country so it might show as [country]. Just insert the name of the country you are referring to.

Formatting and pasting content into documents

When you copy text from one place and paste it into another, from a website into Word, from an AI conversation into Atticus, or from an email into a document, you are often bringing hidden formatting with it that you cannot immediately see. This invisible formatting includes things like font type, font size, line spacing, colour, and paragraph style, all of which were applied in the original source and travel along with the text when it is copied.

In a document you have carefully formatted to a consistent style, pasted text that carries its own formatting can immediately disrupt the appearance of your work. A paragraph may suddenly appear in a different font, a heading may change size, or spacing between lines may shift without any obvious reason. This is one of the most common frustrations people experience when working with word processors and book formatting software such as Word or Atticus.

Why this happens

Every piece of text exists inside a container of formatting instructions. When you copy that text, those instructions come with it. The document you paste into may apply its own styles on top, resulting in a conflict between the two sets of instructions, which is why pasted content can behave unpredictably and prove difficult to correct by highlighting and reformatting manually.

How to paste without formatting

The solution is to paste the text as plain text only, stripping away all the original formatting so that the text arrives clean and takes on the styling of the document it is being pasted into.

There are several ways to do this depending on your device and software:

Windows: Instead of the usual Ctrl+V to paste, use Ctrl+Shift+V. This opens a Paste Menu option, then select Plain Text or Unformatted Text.

Mac: Use Command+Shift+Option+V in most applications, or look for Paste and Match Style in the Edit menu. This pastes the text in the style of the surrounding document rather than the source it came from.

Microsoft Word: Click the small clipboard icon that appears after copying and select Paste without Formatting.

Google Docs: Use Ctrl+Shift+V on Windows or Command+Shift+V on Mac to paste without formatting.

Downloading Apps or adding Extensions

Sometimes you wish to use specialized content or download a programme, install an App, or adding an Extension to your browser, which you have no idea how to manage or access. Simply ask AI how to do this, give it the name of the Extension, the name of your browser, then instruct it to provide you with each step. Work with it together and you will succeed!

When AI gets it wrong

A practical habit worth developing

If you regularly copy content from AI conversations, websites, or emails into your documents, making 'paste without formatting' your default habit rather than the exception, will save considerable time and frustration. Text that arrives clean is far easier to work with than text that arrives carrying invisible instructions fighting against your carefully built document style.

Pasting with formatting is like a new horse arriving already convinced it should be kept in a different yard. Pasting without formatting is like a horse that arrives quietly and settles into whatever routine you already have in place.

Fact-checking, hallucinations, and keeping your reputation safe

Learning to prompt AI well is one skill. Knowing what to do when AI confidently gives you incorrect information is another skill entirely, and it is just as important.

AI systems do not always know when they do not know something. Unlike a person who might say "I am not sure about that one," AI will often produce a confident, well-written, completely plausible answer that is absolutely wrong. This is what is meant by the term hallucination. The AI has not lied to you deliberately. It has generated text that follows a logical pattern based on its training, but that pattern has led it somewhere inaccurate.

For horse people, this counts, because the horse industry contains a great deal of specialist knowledge that AI has learned imperfectly. Bloodlines, breed standards, regulatory requirements, health protocols, competition rules, and the names of specific horses, sires, and dams are all areas where AI can sound authoritative while being quietly wrong.

Publishing incorrect information in your area of expertise is damaging to a reputation that may have taken decades to build. The solution is not to avoid AI, but to develop a habit of verification before anything important goes out.

Pedigrees and bloodline checks

Pedigrees and bloodlines are a common problem. AI may name a sire confidently that does not exist, or place a horse in the wrong generation of a breeding line. Always verify specific pedigree claims against studbook records or trusted breed registries before including them in promotional material or published work.

Breed standards and registration requirements can change, and AI training data has a cutoff date. Requirements you ask about may have been updated since the AI last learned anything. Always check current requirements directly with the relevant breed society or registry.

Health, medication, and veterinary information carry the highest risk. AI can produce plausible-sounding but incorrect dosage information, outdated treatment protocols, or inaccurate descriptions of conditions. Use AI to help you understand and communicate, not to diagnose or prescribe.

Competition rules and regulations change regularly. AI may describe rules that were accurate at some point but have since been revised. Always check current rule books directly.

Historical facts about specific horses, events, and people in the industry are another area where AI can be confidently wrong. If you are writing about the history of a breed, a famous stallion, or a significant event, verify the specific details against primary sources.

Even what I have written in this book will probably be outdated by the time you read it!

A verification habit

Before publishing anything that contains specific facts, names, dates, figures, or technical claims generated by AI, read through with this question in mind: do I know this to be true from my own knowledge or a reliable source, or am I taking AI's word for it?

If you are taking AI's word for a fact that matters, check it. This does not need to be laborious. For most content AI produces, particularly communication templates, social media drafts, and general writing assistance, verification is not necessary because the content does not contain specific factual claims. It is specific claims that need checking, not every word AI produces.

Using AI to check AI

One technique is to ask AI to identify its own uncertain ground. After AI produces a response, you can follow up with:

> *Please review what you just wrote and flag any specific facts, names, dates, figures, or claims that I should verify independently before publishing. Be honest about where you are uncertain.*

This does not catch everything, but it does prompt AI to flag the areas most likely to need checking, and it often surfaces genuine uncertainty that did not show in the original response.

Advanced prompt:

Theme: Fact-checking AI output before publishing

Suited to: anyone using AI to produce content that will be published, shared, or used professionally

Act as a careful editorial assistant with knowledge of the horse industry. I am going to share a piece of content that was written with AI assistance, and I want you to review it specifically for accuracy risks before I publish it.

Please do the following:

Read through the content and identify every specific claim that contains a name, date, figure, pedigree detail, breed standard, health or veterinary reference, competition rule, or historical fact.

For each one, tell me whether you are confident it is accurate, uncertain about it, or cannot verify it from your training data. Flag anything that you believe may have changed since your training data cutoff and that I should check against a current source.

Suggest the most appropriate source I should use to verify each flagged item, for example the relevant breed registry, a veterinary reference, or the current rule book for a specific discipline.

Once you have flagged everything, give me an overall assessment of how much of this content I should verify before publishing and why.

Here is the content: [paste your AI generated content here]

The next chapter takes this further by exploring how to make AI sound like you, not like a marketing department.

Chapter 6

Sounding like yourself

Will people know I used AI?

One of the first concerns you might have about artificial intelligence is not the technology itself. It is the writing.

Will people know I used AI? Will it sound fake? Am I cheating? What about plagiarism? And why does everything suddenly contain strange dashes, American spelling, and phrases that sound as though they were written by an over-enthusiastic marketing department? Horse people are not worried about neural networks. They are worried about reputation, trust, sounding foolish, accidentally plagiarising something, or ending up with content that sounds like a corporate robot wrote it.

These are sensible questions, particularly in the horse world where reputation matters. People buy horses from sellers they trust. They send horses to trainers they trust. They listen to instructors, veterinarians, farriers, and mentors they trust. Authenticity still carries the most weight.

That is why you may feel uncomfortable when you first encounter AI-generated writing. Sometimes it sounds polished but oddly hollow or robotic. Sometimes it sounds as though somebody fed a corporate brochure into a blender and poured it over a perfectly ordinary sentence.

The solution is surprisingly simple.

AI is a tool, not an author. If you ask it to help draft a newsletter, organise ideas for a website, improve a social media post, or create an educational handout, it is assisting you. The experience, judgement, knowledge, and responsibility remain *yours*. That is no different from using a word processor, a spelling checker, a calculator, or a search engine you are already familiar with. The tool helps. You make the decisions.

Plagiarism

One concern that deserves a little more discussion is plagiarism. The word is often used whenever AI enters the conversation, but plagiarism has a specific meaning. It involves presenting somebody else's work, words, ideas, or intellectual property as your own *without proper acknowledgement.*

Using AI is not automatically plagiarism. However, it is still important to use common sense.

If you are quoting somebody, acknowledge them. If you are using research, statistics, or specialist information, verify it. If you are producing academic work, professional reports, or material that requires references, follow the rules that apply to that situation. AI can help organise information, but it should never be used as an excuse to abandon good practice.

Another issue people notice quickly is that AI often develops certain writing habits. It can become repetitive. It sometimes uses dramatic language where none is required. It tends to sound overly enthusiastic, and for reasons that remain something of a mystery, many AI systems have developed a deep affection for em dashes (which look like this —)

The good news is that these patterns are not permanent. Instead you can tell the AI exactly what you want.

Ask it to use Australian English. Ask it to avoid em dashes. Ask it to use sentence-case headings. Ask it to avoid corporate jargon, marketing language, exaggerated claims, and unnecessary complexity, *if that is what you want!*

The more guidance you provide, the better the results.

Show AI examples of your own writing. This might be a newsletter, an article, a website page, educational material, or even a collection of social media posts. Once the system has examples, ask it to analyse your writing style and use a similar tone in future responses.

This often produces a dramatic improvement because the writing begins to reflect your vocabulary, sentence structure, and communication style rather than a generic internet voice. Remove or add in preferences to fine-tune it.

Ultimately, the goal is to create writing that sounds genuine. The strongest use of artificial intelligence is allowing it to become *your* voice. Use it to help express your voice more clearly.

What it looks like in practice

Here is a short example. The first version is what AI produces without guidance; technically correct, perfectly hollow. The second is the same information after the writer has put themselves back into it.

Both describe the same stallion. One sounds like a brochure. One sounds like a person.

Version 1: Unedited AI output

Introducing Cefngarw Caradog, an exceptional Welsh Cob stallion combining outstanding movement, correct conformation, and exceptional temperament. Standing at 15.1hh, Caradog offers breeders a rare opportunity to access proven bloodlines with demonstrated success across multiple disciplines. His progeny consistently demonstrate the versatility, trainability, and presence that modern breeders demand. Limited bookings available for the upcoming season.

That is not wrong, but it sounds like every other stallion advertisement written in the last twenty years. "Exceptional" appears twice in four sentences. "Rare opportunity" is doing a lot of heavy lifting. Nothing in it tells you anything specific about this horse.

Version 2: After the writer edits it back to their voice

Caradog has been standing here for six seasons now and I still enjoy watching him move. He is honest, straightforward to handle, and he throws that same steadiness reliably into his foal, which is really important when you are breeding horses that ordinary people will live with. He suits mares with good bone and a sensible brain. We keep bookings limited because we want to know where his foals are going.

Completely different feeling. The second version tells you something real about the breeder's values, the stallion's character, and how this program operates. The goal is to use AI for the scaffolding, and then put yourself back in. In this instance, I would merge both together, and refine it, using AI.

That is what AI cannot do on its own. It can organise information efficiently. It cannot replace the specific, earned, personal knowledge that makes one breeder's voice different from every other.

> *Act as an editor, plagiarism checker, and writing coach. Please review the following text [insert] and revise it so it sounds natural, authentic, and human.*
>
> *Requirements:*
>
> - *Use [country] English spelling and grammar.*
> - *Use sentence case headings.*
> - *No em dashes.*
> - *Avoid corporate jargon, marketing language, clichés, and exaggerated claims.*

- *Preserve my original meaning, experience, and voice.*
- *Improve readability, flow, and sentence variety.*
- *Identify any passages that may be too similar to published sources and suggest ways to make them more original.*
- *Flag any statements that require attribution, citation, or fact checking.*
- *Ensure the final version sounds as though it was written by an experienced horse person (me) rather than generated by AI.*
- *After editing, provide a summary of any plagiarism concerns, attribution issues, or significant changes you made.*

AI safety and common sense

A simple rule helps. Ask yourself:

- Does this make sense?
- Can I verify it?
- Would I be comfortable acting on this information?
- If the answer is no, slow down.

You do not believe every piece of advice heard around a stable just because it is delivered confidently. You question it, compare it, test it, and use common sense. Apply the same approach here.

Act as a critical thinking assistant. Focus on balanced analysis rather than certainty. Please review the following information and identify:

- assumptions
- possible errors
- areas requiring verification
- questions I should investigate further
- risks of acting on this without checking it

A Caution About Safety

The horse industry has always been a community. People help each other. They share knowledge, support one another through difficult times, and genuinely care about each other's success. That sense of belonging is one of the most beautiful things about horses.

But something has shifted quietly over the past decade, and what happened at the South Point Casino in Las Vegas in May 2026 exposed a dangerous truth we can no longer ignore.

A seventeen-year-old girl who had been following barrel racer Ariel Phillips online for months showed up at the stables after midnight. She had constructed an entire friendship in her head with someone who did not know she existed. When she encountered Ariel in person and discovered she was just a stranger, something inside her broke. She pulled out a knife and stabbed three horses: Detail, Sully, and Rocket. Then she washed the blood away, attempting to insert herself into the role of rescuer.

This is not random cruelty. This is a predictable psychological crisis that our industry has been building towards for years.

An equestrian's horse is not a piece of equipment. It is a partner. A horse is your identity, your financial stability, your purpose, and in many cases, your entire sense of self-worth. For a para-social stalker, attacking the horse is the most devastating way possible to inflict permanent emotional damage on the creator.

The Structural Problem

To build sponsorships and keep audiences engaged, equestrian creators post their exact show schedules, locations, stall numbers, and horse names. An obsessed fan does not need to break into a mansion. They just drive to the arena, walk past an unmonitored gate at eleven at night, and stand outside a defenceless animal's stall.

What Needs to Change

If you are an equestrian content creator your digital audience is a community, but they are not your friends. Protecting your boundaries, hiding your stall numbers, obscuring your real-time location, and keeping your private life private is not ungrateful. It is necessary for your safety and your horse's survival.

You can get the full story from this video on YouTube
https://www.youtube.com/watch?v=O3XWEukbrZQ

Your first result is rarely your best result

AI occasionally misunderstands instructions, invents strange wording, misses context, creates awkward formatting, produces emotional nonsense, or decides a horse apparently has six legs in a promotional image.

This is normal. And understanding this properly is extremely important. Because good AI use is not about pressing a button once and blindly accepting whatever appears. Good AI use is collaborative, particularly in the horse world where realism matters enormously.

Your first result is rarely your best result. This is something beginners need to understand immediately. Many people try AI once, receive a mediocre answer, and decide it was disappointing. Usually the issue is not the technology. It is the prompt. AI improves dramatically when instructions become clearer. Think of prompting like briefing a new stable hand. If you merely say: "Feed the horses." You may get wildly different outcomes.

But if you say: "Feed the broodmares first, separate the yearlings, soak the beet pulp, and monitor the chestnut mare carefully because she is particularly bossy," the instructions are clear.

> Weak: *Write horse ad for my gelding.*
>
> Much stronger: *Write a calm and professional sales advertisement for a five year old Welsh Cob gelding suitable for an experienced amateur rider. Focus on temperament, handling manners, consistency, and realistic expectations without exaggeration.*

The second prompt gives tone, audience, and horse type, boundaries and communication style.

That creates much stronger results. This chapter may actually be one of the most important in the entire book. AI-generated writing should NEVER be copied and pasted blindly. Editing still matters enormously. AI produces beautiful clear wording. Sometimes it produces sentences sounding like: "A majestic equine companion optimised for synergistic rider alignment," which sounds less like a horse and more like an expensive yoga retreat.

You still need to check accuracy, tone, grammar, horse realism, breed correctness, safety issues, emotional tone, awkward wording.

How to improve weak AI responses

You can refine responses repeatedly. This is where AI becomes much more useful. You shape the result gradually.

For example:

Rewrite this to sound more natural and less exaggerated.

Shorten this advertisement and make it sound more conversational.

Remove repetitive wording and improve clarity for beginner horse buyers.

Please challenge the assumptions in my current understanding of this topic and identify any blind spots I may have overlooked.

Why downloads sometimes fail

This deserves mentioning because technology occasionally behaves like a nervous young horse encountering a plastic bag in crosswinds.

Downloads fail sometimes. Formatting breaks. Files corrupt. Images disappear. Fonts shift unexpectedly. Particularly when files become large, formatting becomes complicated, or multiple platforms interact together.

This is normal. And importantly it is usually fixable. Simple ways to reduce download problems are by using standard fonts, clear headings, plain layouts, and basic file structures.

Avoid overly-complex formatting, excessive graphics, multiple font combinations, and extremely large image files.

Simple documents usually transfer more reliably between Word, Canva, Atticus, PDF systems, Website downloads and the like.

AI image creation

Artificial intelligence can now generate book covers, social media graphics, promotional flyers, stable signs, educational graphics and advertising concepts in an image format. But AI can seriously distort horses.

AI image systems do not "understand" horses like humans do. They generate images by predicting visual patterns from enormous amounts of training imagery. Which explains many terrifying AI horse images currently roaming the internet. Some look like emotionally unstable centaurs; others completely fabricated, like the mare with triplets.

That means:

- Common horse shapes usually work better
- Complicated angles create error
- Fast movement increases distortions

- Multiple horses increase confusion
- Detailed tack increases mistakes

How to improve horse image prompts

The clearer and simpler the image prompt, the better the result usually becomes. Standing horses usually generate more accurately than fast action shots.

Weak: *Create horse image of horse standing near stable*

Better: *Create a realistic buckskin Quarter Horse standing quietly in natural afternoon light beside a timber stable, with a saddle resting over the doorway. Clean background. Correct anatomy. Correctly fitted halter. Relaxed expression. No dramatic effects.*

Create realistic galloping horse with correct anatomy and correct leg positioning, natural movement, accurate hoof structure, realistic bridle and reins, and proportionate bareback western rider in position, background American rural scene. No distorted legs, extra limbs, warped tack, unrealistic posture, or exaggerated movement.

You can also ask AI to:

- Simplify or change colours and background
- Reduce motion
- Change a profile view
- Change a bridle to a halter
- Change to a historical or fictional setting.

AI can add text to promotional images

Create a clean promotional flyer for a buckskin Saddlebred stallion at stud with elegant typography, realistic horse presentation, including stud fee, bloodlines, breeding information, temperament highlights, service information, location, website addresses, and contact details. [insert photo and details.]

These graphics can then become Facebook advertisements, handout flyers, website graphics, magazine concepts, and brochures.

The strongest horse people are still human

Your personality is YOUR VOICE!

The most trusted horse people still sound like real people because they reflect genuine values, communicate warmly, share real experiences, and show real horses. They reflect the good and the bad. They tell their personal stories, the mistakes, the lessons, the hardships, the amazing experiences.

And that combination of expertise and authenticity is becoming increasingly powerful online. Because in a world filled with noise, calm and trustworthy communication stands out immediately.

Final thoughts

AI is not magic. It is a tool, sometimes brilliant, sometimes annoying. Usually somewhere in between.

The horse people who benefit most will not necessarily be the most technical people. They will probably be the people who stay curious, learn gradually, edit carefully, communicate clearly, protect authenticity, and use technology thoughtfully.

You simply need to understand enough to make AI work for you, while keeping your own knowledge, judgement and horsemanship at the centre of every decision.

> *Act as an AI quality control assistant. Please review the following AI generated content and identify factual errors, unrealistic claims, exaggerated language, horse industry concerns, safety and welfare concerns, marketing risks, and credibility issues. Then suggest improvements while maintaining honesty and realism.*
>
> *Act as a professional research assistant. Please help me investigate the following horse related topic [insert topic]. Provide an overview, key concepts, important questions, areas of agreement, areas of disagreement, and useful sources for further investigation. Clearly separate established facts from opinion or speculation.*

***

Build your personal AI action plan

Step 1. Identify your biggest frustration

Artificial intelligence works best when attached to real problems. Ask yourself:

- What task do I dislike doing?
- What takes too much time?
- What do I repeat constantly?
- What causes unnecessary frustration?

Examples might include:

- answering enquiries
- writing newsletters
- organising records
- managing bookings
- creating educational material
- updating websites
- preparing documentation
- organising photographs

Start with a genuine problem. Not a fashionable tool.

Step 2: Choose one area only

Don't try to improve everything at once. This usually creates overwhelm. Choose one area. For example:

- Communication.
- Marketing.
- Organisation.
- Education.

- Administration.
- Scheduling.
- Record keeping.

Step 3: Define a small goal

Good goals are specific. Small goals create momentum.

Examples:

- Create a client welcome pack.
- Organise breeding records.
- Write one newsletter.
- Build a frequently asked questions page.
- Create a simple website.
- Develop a lesson information guide.

Step 4: Experiment

Give yourself permission to experiment. Not everything will work perfectly. Some prompts will be unhelpful. Some ideas will be discarded. Some tools will not suit your needs. That is normal.

Step 5: Evaluate results

After completing a project, ask:

- Did this save time?
- Did this improve communication?
- Did this reduce stress?
- Did this help clients?
- Did this improve organisation?
- Would I use this approach again?

The answers provide valuable guidance.

Step 6: Build gradually

Once one system works, improve another. Small improvements accumulate. A newsletter here, a website update there. Better records. Clearer communication. More organised information. Over time these changes become significant.

Your first year

Imagine making only one meaningful improvement each month. Twelve improvements. Twelve better systems. Twelve new resources. Twelve areas of reduced friction. That is substantial progress. And it feels much more achievable than attempting everything immediately.

A final thought

The future rarely changes because somebody learns something new. The future changes when somebody does something with what they have learned.

Personal action plan worksheet

Complete this after reading the action plan steps above.

- My biggest frustration with running my horse business is:
- The one area I want to improve first is:
- My first AI project will be:
- I will complete it by:
- The result I hope to achieve is:
- The prompt I will use to get started is:

Start with one task. Do it imperfectly. Adjust. Try again.

PART II

Showing Up Online

The constant grab for attention quickly turns to overwhelm and chaos.

Chapter 7

AI for social media

The never-ending demand for attention

For many breeders, trainers, riding instructors, and horse businesses, social media can feel like a never-ending demand for attention layered on top of an already busy life. The horses still need feeding. The fences still break. The vet arrives late. And meanwhile, the internet wants educational posts, video updates, polished captions, engaging stories, professional branding, and endless consistency.

This is one of the areas where AI can genuinely reduce stress without sacrificing authenticity. Not by replacing your voice, but by helping you organise and communicate your knowledge more effectively.

You have valuable content

The biggest misconception about social media is that you need a constant stream of new ideas. In reality, most horse people already possess an enormous amount of possible content. They simply do not recognise it as such.

A breeder explaining foal-watch routine, a trainer discussing re-training, a stallion owner discussing temperament, a riding instructor explaining groundwork, a farrier talking about hoof balance, are all providing information that others genuinely want to learn.

The difference between social media and websites

Although people often group them together, social media and websites serve very different purposes.

Social media is fast moving, conversational, frequently updated, and immediate. A website is stable, structured, professional, searchable, and designed for long term use.

One way to think about the difference is this:

- Social media is your ongoing conversation.
- Your website is your permanent business home.

Both matter, but they perform different roles.

Every horse business website should include a clear homepage, an About page, contact details, information about services or horses available, real photographs, business values or breeding philosophy, basic navigation, and mobile friendly design.

Most importantly, it should sound like you, not a corporate marketing department attempting to sell "premium equine optimisation experiences."

AI can help create website content quickly and professionally.

Write a calm and practical homepage introduction for a small horse breeding farm focused on temperament, trainability, and family horses.

Help me write an About page explaining our agistment philosophy and horse management values in a warm, conversational tone.

Your knowledge has value

Websites also provide a home for educational resources. Many horse people have accumulated enough experience to create:

- educational downloads
- tack measurement templates
- buyer education packs
- prompt libraries
- young horse training journals
- checklists and workbooks
- workshops

Unlike social media posts, these resources can be organised permanently, delivered automatically, and used repeatedly. That creates value not only for your audience, but also for your service.

Educational content builds trust

Educational content positions breeders, trainers, and horse professionals as thoughtful, knowledgeable, and trustworthy without constantly sounding promotional. It is one of the most effective forms of communication. AI can help structure topics into clear, readable content.

Topics might include feeding advice, young horse development, temperament, training observations, foaling preparation, rider suitability, stable management, welfare considerations, or breeding decisions. Content like this often builds trust more effectively than endless sales messages.

> *Help me write a calm educational Facebook post explaining why feeding senior horses appropriately matters.*
>
> *Write a short educational post explaining why young horses need time to mature mentally as well as physically.*

AI is excellent for content ideas

AI is remarkably good at helping generate ideas. The expertise remains yours. AI helps organise it into usable topics and content plans.

> *Give me 20 educational Facebook post ideas for a horse breeding farm focused on temperament, trainability, and responsible ownership.*
>
> *Suggest social media content ideas for a Thoroughbred breeding programme during foaling season.*

Tell the AI how you want the content to sound. The more specific you are about tone, the more natural the result becomes.

> *Write this Facebook post in a warm, practical, conversational tone suitable for experienced horse people. Avoid corporate or overly dramatic language.*
>
> *Make this educational post about [topic] sound calm, grounded, and like a mentor, rather than pushy or sales focused.*

One piece of content can become many

AI can help restructure the same information for multiple uses. This dramatically reduces the amount of time spent creating content from scratch.

A single article about your foaling season might become:

- Facebook posts

- educational captions
- newsletter content
- website articles
- buyer handouts
- frequently asked questions
- email updates

Turn this article into five short educational Facebook posts.

Summarise this breeding article into three social media captions suitable for horse owners.

Newsletters matter more than most people think

You instinctively know you want to connect more consistently with buyers, clients, members, or followers. The challenge is maintaining that connection when everything else competes for attention.

AI can help draft newsletters, event updates, breeding announcements, foaling reports, educational emails, sale notifications, when horses are undergoing rehabilitation, or students searching for work experience. Occasionally internet forums decide to become emotionally invested in something deeply ridiculous or divisive that you need to be aware of. You might ask AI to prepare a response.

The goal is not perfection. The goal is maintaining communication without facing a blank screen every month.

Consistency is one of the biggest challenges not because you lack ideas, but because life constantly intervenes. A little structure often removes a surprising amount of mental overwhelm.

Help me write a monthly newsletter update for our club, including news, educational content, and upcoming clinics. Ask me questions for topics to include, as we go along.

Example: Annual Christmas newsletter

The following is an example of an AI-assisted annual newsletter for a Welsh Cob breeding program. This was drafted in about fifteen minutes, from a set of brief notes, then edited to restore the breeder's voice. It is a working example, not a template. Adjust every detail to match your own program, horses, and personality.

Seasons greetings from [Stud Name and Logo]

[photo — herd shot or property image, festive if available]

Dear friends, clients, and fellow horse people,
As another year draws to a close, we wanted to take a moment to share some news from the property, say thank you to everyone who has been part of this year, and let you know what is coming up in [year].
It has been a full year. A good one.

This year's foals

[photo — individual or group foal photos]

We welcomed four Welsh Cob foals to the property this season, and we are genuinely pleased with how they have all developed. Two colts and two fillies, each one different in personality, each one very much carrying the temperament and bone we breed for.

All four are growing steadily and are being handled quietly as part of their daily routine. We will share more photos and updates as they develop through summer. If you have enquiries about future availability, now is a good time to get in touch.

[photo — foal close-up or handling photo]

Student achievements

We are proud to share two particular highlights from our students this year.

[Student name] competed at [event/level] in [month] and achieved [result]. It has been a pleasure watching her progress over the past [time-frame], and this result reflects the consistent, quiet work she has put in.

[Student name] passed [qualification/milestone] earlier this year. This has been a long-term goal and we are delighted to have been part of that journey.

Well done to both. These things do not happen without real effort.

Did you know?

Welsh cobs are one of the few breeds that has remained largely unchanged in type/character for centuries. The qualities that made them essential working horses (soundness, stamina, good bone, a sensible brain) are the same qualities that make them such reliable family performance horses today.

When we select for temperament above all else, we are continuing a tradition that goes back a very long way. That is not sentiment. It is good breeding practice.

Property open day — save the date

[photo — property, arena, or stable yard image or scenes from last years Open Day]
We are holding an open day at the property on [DATE], and you are warmly invited.

The day will include an opportunity to meet the horses, see the foals, watch a short handling demonstration, and ask questions in a relaxed setting. There will be [refreshments / morning tea / whatever you plan to offer].

Numbers are limited to keep the day relaxed and unhurried. We want people to have genuine time with the horses, not a crowded viewing.

To book your place: [contact details / booking link / email address]

Places will be confirmed by [date]. We look forward to seeing you here.

Thank you

To everyone who has supported the programme this year, through purchases, referrals, kind words, or simply following along, thank you. The horse industry runs on trust and reputation, and we do not take either lightly.

Wishing you and your horses a safe, peaceful, and well-deserved rest over the holiday season.

Warmly,

[Your name and position]

[Stud name] | [Website] | [Phone] | [Email]

[photo — closing image, herd at pasture or seasonal photo]

You do not need to post constantly

Be reassured, you do not need to become a full-time content creator. You do not need to post six times a day. You do not need dramatic videos filmed at sunrise while delivering motivational speeches beside an arena fence.

Consistency matters far more than intensity. This is especially true in the horse world, where reputation still matters and credibility is built over time.

Create a simple weekly social media content plan for a month's post ideas, with short caption suggestions, for a horse stud focused on education, breeding philosophy, and horses for sale.

Act as a horse industry content strategist. Create a one-month social media content plan for my horse programme for [insert social media names you use].

Include educational posts, engagement posts, behind the scenes content, client focused content, horse welfare topics, and seasonal opportunities.

For each post include a headline, caption, and call to action.

The Snipping Tool

Most people have been taking screen captures for years without thinking of them as a technology skill at all.

On a Windows computer, the Snipping Tool is built in and worth getting to know properly. You can find it by typing Snipping Tool into your search bar, capture any portion of your screen instantly by dragging a box around exactly what you want.

It allows you to capture just the relevant section rather than the entire screen, which is far more practical when you only need a specific paragraph, a pedigree section, or a feed label.

The reason this matters for AI users is significant. You can snip a veterinary report, a breed standard, a competition result, or even a handwritten note and ask AI to analyse, summarise, translate into plain language, or respond to it.

If AI struggles with very small text, snip a smaller, closer section and try again.

Snipping tools are excellent for social media work. You can capture a post that performed well, an image, a competitor's advertisement that caught your attention, or a comment thread worth responding to, and ask AI to analyse what worked, suggest a response, or help you create something similar in your own voice.

One important note on quality: snipping tools do resize images. However, if the original image is high enough resolution, the snipped result is often still suitable for print purposes. I have found this to be particularly useful when working with historical photographs, where the original may exist only as a scan or digital file, or you wish to crop out some background to increase the size of a section of the image.

Snipping a clean section from a high resolution original will generally give you a better result than trying to work with a poor quality copy of the whole image. As always, zoom in on your snipped image before committing to print to check that the detail and clarity are holding up.

Advanced prompt:

Theme: 90-day seasonal content plan

Suited to: studs, training operations, single-owner horse businesses, equine practitioners

Act as a social media strategist with deep knowledge of the horse industry. I run [describe your business, e.g. Riding Pony stud / a dressage training yard / farriery / dental practice] and I want to create a 90-day social media content plan starting [insert month]. Write in a tone that is warm, professional, and honest, not corporate.

My audience is [describe: e.g. mare owners, competition riders, horse-owning families in regional areas]. My main platform is [Facebook / Instagram / both].

Build me a 90-day plan that rotates across four content types:
1. educational posts that demonstrate my expertise,

2. behind-the-scenes content that shows real life on the property,

3. promotional posts for [your key seasonal focus, e.g. stallion service fees / upcoming availability / competition results], and

4. community posts that invite engagement or share client stories.

Time the plan around [insert your seasonal focus, e.g. breeding season / competition season / foaling / winter]. Give me one post idea per week at minimum, with a suggested caption angle, the best time to post, and whether the post should include a photo, video, or graphic.

For further ideas on using AI for social media content, this practical article is worth reading:

https://djangoandy.com/2026/02/19/using-ai-to-create-social-posts-for-your-business/

Chapter 8

Authenticity and Reality

Real photos, honest marketing

The horse industry has a complicated relationship with presentation. In the show world, one practices presentation for years to best cover a horse's faults and emphasise its good points. This is a form of advertising; in this case to present to a judge and spectators for evaluation. You only get one chance to show off your horse before the judge looks at another.

Good advertising is no different. Clear photographs, polished branding, and thoughtful marketing help horses gain visibility and help breeders communicate more effectively.

On the other hand, the industry has become increasingly saturated with unrealistic imagery, over editing, filtered perfection, and now, AI generated horse images that often look impressive but bear little resemblance to reality.

That creates a serious problem. Because trust matters deeply in the horse world and trust begins with integrity.

Why authenticity matters

While AI can absolutely support communication and branding, it should never push you toward fake presentation or manufactured perfection.

Experienced horse people recognise exaggerated advertising almost immediately. The impossibly polished sales photos, the dramatic wording, the suspiciously perfect movement shots, the advertisements where every horse somehow appears suitable for nervous beginners, dressage competition, trail riding, and mounted games, while simultaneously solving your emotional trauma.

Increasingly, people are craving something quite different online. Simple, direct, genuine, that feels like they are having a conversation within a community with a real person, not polished fantasy.

Images and photos

Real photos build confidence. This surprises people. They assume buyers only respond to highly Polished glamorous images or synthetic brochures.

Buyers connect far more strongly with photos that feel genuine. A horse standing naturally in good light, a sensitive horse being handled calmly, a quiet moment during grooming, a horse interacting softly with people. Such images communicate far more than dramatic editing ever could. Audiences want to feel that the horses seem well cared for. They are looking for reassurance.

Because buyers are subconsciously asking:

- Does this horse look settled?
- Does this environment feel calm?
- Does this breeder appear organised and genuine?
- Do these horses look well-managed?

Photos instantly communicate enormous emotional information. AI image generators are now capable of creating horses that appear visually impressive at first glance but are entirely fictional. These horses may have impossible movement or impossible lighting.

If fake images are used in advertising, breeding promotion, or sales material, misleading visual presentation when identified by an audience, damages trust. And once that disappears, reputations can become difficult to rebuild.

This is why this guide strongly recommends using real horses, real photographs, and real environments. AI should support communication around your horses, not replace the horses themselves.

So what is good photography?

Good photography is about clarity, not perfection. That means good lighting, clean presentation, clear angles, minimal distractions, realistic representation, and calm handling. Buyers prefer real, slightly imperfect images, over heavily manipulated photographs that create unrealistic expectations. Photos showing horses being managed quietly, standing calmly tied, or interacting naturally with people, can communicate temperament beautifully.

However some improvements can make an enormous differences. Only do the following if you cannot obtain a good photo:

Ask AI to:

- Use natural light: soften with early morning or late afternoon light. This is usually softer and more flattering than harsh midday sunlight.

- Keep backgrounds simple: erase messy yards, overflowing feed bins, random machinery, loose dogs sprinting through the frame, or a handler (if required).
- Show horses realistically: Do not attempt to disguise flaws through strange angles or excessive editing.

Never alter the horse itself. The buyer will eventually meet the horse in person.

Copyright and photo permissions

If you did not take the photograph yourself, you do not automatically have the right to use it. This is not a grey area. Using a photograph you found online, on Google Images, Facebook, another breeder's website, a breed society page, or anywhere else, without the explicit permission of the photographer or owner is a breach of copyright. It does not matter whether the image has a watermark or not. It does not matter whether you credit the source. Using it without permission is illegal.

The consequences are real. Photographers and image libraries actively search for un-authorised uses of their work online. A formal copyright claim can result in a demand for payment, removal of the image, or legal action. This has happened to more than one consumer. It will continue to happen.

The rule is simple: only use photographs you have taken yourself, or photographs where you have written permission from the photographer to use them for commercial purposes. If you hire a photographer, clarify ownership of the images in writing before the shoot. Many professional photographers retain copyright even when you pay for a session.

Free image sites such as Unsplash and Pexels offer photographs licensed for commercial use at no cost. These are a legitimate option when you need a generic image. For anything representing your specific horses, your property, or your program, use your own photographs. There is no substitute.

Building on the image

This is where AI becomes extremely useful without crossing ethical boundaries. AI can help generate captions, wording overlays, flyer concepts, advertisement layouts, educational descriptions, and social media text.

Use photos of daily stable life, real horses, foals growing up naturally, horses interacting calmly, or genuine behind the scenes moments. For example:

> *Help me write professional wording overlays for a stallion advertisement using this image [insert real image]. Include stud fee, location, service options, and*

temperament highlights as listed [here].

Write an inviting, warm caption for a sales photo of my young Quarter Horse gelding. Focus on temperament and reliability.

Unrealistic expectations create poor matches. Honest presentation protects horses. Technology becomes dangerous when people start using it to manufacture illusion. No amount of technology replaces integrity.

Act as an ethical horse industry marketing consultant. Please review the following [advertisement, website page, social media content, or promotional material]. Identify anything exaggerated, unrealistic claims, trust issues, opportunities for greater authenticity, and areas where clearer communication would help.

Your next advertisement has one job; to describe the horse you have, not the horse you wish you were selling.

If something doesn't work

Images are readily made on Chat GPT, and it is very good at editing photos provided, but it does have its limitations. For example when darkening something for contrast, it may also darken natural skin tones. You have to tell it quite clearly, *do not change anything else.*

You have two practical options if something doesn't work:

Option 1: Accept it and fix it in a design tool: For example, download the image as-is and open it in Canva. Use the image editor's Brightness, Warmth, and Tint sliders to warm the overall image slightly. It's a blunt tool but it will shift the whole image warmer together, which may be enough given a horse and rider are both affected by the same cool tone issue.

Option 2: Use a photo editor: There is a free Photoshop-like tool called Photopea, that can allow more editing options than Canva. If you or someone you know has access to Photoshop, Lightroom or even the app Snapseed on a phone, the selective colour or brush tool can warm just a horse and rider without touching the background. Snapseed in particular has a "Selective" tool that lets you tap on one area and adjust warmth/saturation there only. It's free and takes about five minutes

Advanced prompt:

Theme: Voice audit and ghost-writing blueprint

Suited to: any horse professional wanting consistent, personal content

Act as an editorial consultant specialising in authentic voice development. I am going to paste three examples of writing I have done myself , a social media

caption, a horse advertisement, and an email to a client. Read all three carefully before doing anything else. [Paste your three examples here]

Now do the following: First, identify the specific patterns that define my natural writing voice, my sentence length, the words I tend to use, the way I open and close paragraphs, my level of formality, and any recurring phrases or rhythms.

Second, write a one-paragraph summary of my voice that I can use as a briefing note for future AI sessions.

Third, use that voice to write [choose one: a stallion profile / a post about a recent foal / a response to a client enquiry about availability], something I could publish today without editing it to sound like me, because it already does.

Image size and quality

Image size and quality matter more than most people realise, and choosing the wrong one can result in a photograph that looks sharp on a phone screen but blurry and unprofessional when printed in a book or magazine.

Digital images are measured in pixels, which are the tiny dots of colour that make up a photograph. The more pixels an image contains, the more detail it holds. You will often see this expressed as dimensions such as 1920 x 1080, meaning 1920 pixels wide and 1080 pixels tall. The total number of pixels is also referred to as megapixels. A 12 megapixel image contains approximately 12 million pixels.

Resolution, measured in DPI, which stands for dots per inch, describes how densely those pixels are packed when the image is printed or displayed. The higher the DPI, the sharper and more detailed the result.

What size do you need? As a practical guide:

- **72 DPI** is standard for screens, website images, and social media posts and email. It looks fine on a device but will appear soft and pixelated if printed at any reasonable size.

- **150 DPI** is acceptable for basic printed documents such as basic home printing, simple flyers or digital newsletters where fine detail is less critical.

- **300 DPI** (high resolution) is the professional standard for print such as books, magazines, professional brochures, and commercial printing of any kind. Always aim for 300 DPI if your work is going to print.

How to check the size of an image you already have

- On a Windows computer, right-click the image file, select Properties, then open the Details tab. You will see the dimensions in pixels and the resolution in DPI listed there.

- On a Mac, open the image in Preview, then go to Tools and select Show Inspector. The resolution and dimensions are displayed clearly.
- On a smartphone, the information is usually found in the Details or Info section of your Photos app, though smartphones measure in pixels rather than DPI.

If you are unsure whether an image is good enough for print, a basic test is to open it on your computer and zoom in to 100 percent. If it looks sharp and clear at that size, it will generally print well. If it starts to look blocky or blurry, it is not suitable for professional print use.

It is tempting to resize a small image to make it bigger, but this does not add pixels that were never there. Stretching a low resolution image only spreads the existing pixels further apart, making the result look worse rather than better.

AI up-scaling tools such as Topaz Gigapixel intelligently fill in missing detail, producing better results than basic resizing. However, AI up-scaling cannot fully replace a high resolution original image. Always start with the highest quality image available.

I asked AI to provide me with a sales advertisement for my book, using images I provided. I was happy with what it gave me.

Chapter 9

First impressions

Branding your services

Some horse people can spend years building knowledge, bloodlines, handling skills, reputation, and experience, then present themselves online in a manner that looks random at best. That may sound harsh. But branding is important, far more than many horse people realise.

Before clients ever visit your property, speak to you about your event, or enquire about your organisation, they are already making decisions subconsciously based on presentation, not just the horses themselves. The business. The atmosphere. The professionalism. The consistency. The feeling. People instantly ask themselves whether this feels trustworthy, organised, professional, calm, and established. Branding answers these questions silently before a single conversation takes place.

In today's horse world, first impressions increasingly happen online. That makes branding incredibly important.

What is branding?

Branding is not simply a logo sitting in the corner of a Facebook page. Some readers will be running very small or informal operations (leisure horse owners, one-person riding lessons) and may feel the branding discussion doesn't apply to them, but any simple, consistent approach counts as branding

Good branding is the overall feeling your presence online creates consistently. It includes your colours, photographs, wording, communication style, fonts, website, social media presence, advertising style, and values. All of these elements work together to create recognition and reputation over time.

Strong branding feels consistent. Consistency builds confidence. Horse people understand this instinctively because horses respond to consistency too. Clear signals create calmness. Confusing signals create tension. Business works very similarly.

The logo

The apparent default is a horse head silhouette, flowing mane graphic, rearing horse, gold script fonts hard to read, which makes most horse logos look identical. There is nothing wrong with horses appearing in horse business branding. But the strongest branding usually reflects identity rather than clichés.

Good branding asks what makes this programme different, what values matter most, and what feeling people should associate with you and your horses.

Your branding should reflect your vision

This is where AI becomes incredibly useful. Not because it creates your identity for you, but because it helps you think more clearly about the identity you already have.

Instead of searching online for horse logo ideas, try asking:

Help me brainstorm branding concepts for horse training, focused on developing athletic horses for polocrosse competitors, emphasizing my experienced background, care and sound training principles. Avoid generic horse head logos.

Help me brainstorm branding concepts for a 100 year old family breeding programme focused on proven versatility, traditional Australian Stock Horse bloodlines and steady, reliable temperaments.

Immediately, the conversation becomes more thoughtful. Now you are exploring heritage, personality, atmosphere, values, and emotional connection. That creates stronger branding.

Your logo does not have to include a horse

This may surprise you. You might use a polocrosse stick and ball in the first one, for example.

Some of the strongest horse branding uses landscapes, trees, fencing lines, property symbols, family initials, abstract movement, heritage references, or regional identity. Good branding is about recognition and emotional association, not stating the obvious.

If everybody already knows you work with horses, the logo does not need to scream it aggressively from the rooftops. Subtlety often can and does create stronger long-term branding.

Colours matter psychologically

Most people choose colours emotionally without realising colours communicate powerful subconscious messages. Earth tones like greens, browns, and soft neutrals often feel grounded, trustworthy, calm, and natural. Navy and charcoal feel professional, established, reliable, and traditional. Warm creams and muted tones feel elegant, timeless, and heritage focused. Blue can feel bright and cheery.

Good branding should still feel appropriate five years from now.

AI can help with colour direction

It can be fun and rewarding to choose *your* colours. AI is wonderfully practical for this. Ask it several times over, or request subtle changes, so that you can experiment until you find the ones that feel just *right*. Your colour palette can then carry over into everything you do in the future, such as brochures, catalogues, sales and event promotions. This gives you direction without needing expensive branding consultants.

> *Suggest branding colours for a Connemara horse breeding business focused on professionalism, show quality horses, and traditional bloodlines.*
>
> *Help me choose a colour palette for a Welsh Cob farm wanting to feel timeless, grounded, and trustworthy.*

Fonts

If buyers cannot easily read your business name, stud fee, or contact details, the branding has already failed in practice. This may sound obvious but it is not. One of the biggest branding mistakes you could make is choosing fonts that become difficult to read.

Fonts communicate personality instantly. Some feel traditional, elegant, modern, luxury focused, friendly, serious, or fun. Some feel like somebody lost a bet. Clarity matters. Remember your logo may become a thumbnail, so it needs to be read at any size.

Do not hide behind your logo

Though you may not wish to show yourself online, interested parties are increasingly seeking connection. They want to know who they are dealing with, what your values are, how horses are handled, and what your atmosphere feels like. This does not require becoming an Influencer, nor performing motivational speeches beside hay bales at sunset.

AI can help write brand statements and taglines

Horse people already know exactly what they value but struggle putting it into words clearly.

Help me write a short brand statement for a Quarter Horse programme focusing on ranch horses, steady temperaments, and practical training methods.

Suggest tagline ideas for a Clydesdale stud focused on traditional bloodlines and versatile roles. Make it simple, memorable, and clear.

Business cards still matter

Even now, horse people still exchange information constantly at shows, clinics, sales, breed events, stud inspections, and competitions. A clean professional business card still creates a strong impression. AI can help structure layouts and wording.

Help me organise wording and layout for a professional horse acupuncture business card focusing on rehabilitation, to include logo, website, contact details, and social media information. Use the following colour palette [insert].

Good branding is not about looking expensive or flashy. It is about feeling consistent in values, communication and presentation. You build your reputation steadily through clarity, professionalism, and reliability. Branding helps you communicate those qualities visually, helping to establish trust over time.

AI should help horse businesses become clearer versions of themselves. Technology should support you, not replace you, and thankfully, no algorithm can replace relationships, and reputation.

Act as a branding consultant specialising in horse businesses. Help me develop a brand identity that reflects my values, horses, target audience, and long-term goals. By asking me specific questions, suggest branding themes, colour palettes, logo concepts, taglines, tone of voice, and visual style.

Advanced prompt:

Theme: Business positioning statement

Suited to: farriers, trainers, equine body workers, independent horse professionals

Act as a brand strategist with experience in trades, business services, and the equestrian industry. I am a [farrier / equine massage therapist / trainer / insert your profession] based in [region]. I have been working in this industry for [X years] and my particular strengths are [describe two or three e.g. starting young horses, working with nervous horses, rehabilitation after injury, competition preparation].

My ideal clients are [describe e.g. serious amateur competitors, breeding operations, horse owners who prioritise welfare over results]. My biggest competitors in my area are [describe the general type, not names [e.g. eventing and hunting stables, high-volume barns that prioritise turnover, cheaper providers who cut corners on welfare,].

Build me the following:
1. A two-sentence positioning statement that clearly explains who I am, who I serve, and what makes my approach different, suitable for my website homepage.

2. A tagline of ten words or fewer.

3. An elevator pitch of four to five sentences I can use when someone asks what I do.

4. Three key messages I should return to consistently across all my communication. Write everything in a tone that reflects expertise without arrogance, and warmth without being vague.

***

First impressions count, so endeavour to ensure that special emotional connections with your horses shine through. That 'feel' demonstrates a part of your life better than any manipulated content. What you post is what 'connects' you with readers. Photo credit: Martin Waddingham

Chapter 10

Horse sales advertisements

The goal is good placement

Most horse sales advertisements fail for one simple reason. They either tell buyers almost nothing, or they attempt to sound so impressive that the horse no longer resembles a real animal.

Some advertisements are painfully vague. Lovely gelding. Quiet. Regretful sale. That tells buyers absolutely nothing useful.

Others read like the horse personally invented brilliance, floats sideways through paddocks softly at sunset while carrying nervous amateurs to reining success.

Neither approach helps horses.

Good advertising is not about hype. It is about clarity. Clear communication creates better matches, and better matches usually create better long-term outcomes for your horses. Many horse problems begin long before a horse ever leaves the property. They begin with unrealistic advertising.

A horse advertised truthfully attracts buyers who are better prepared for reality. That means fewer misunderstandings, better suitability, better communication, less stress, and better long-term outcomes.

That means explaining temperament, training level, suitability, manners, management requirements, current experience, strengths, and limitations.

Horses pay the price for poor advertising far more often than people realise.

The struggle with writing advertisements

You know your horses extremely well. The problem is translating that knowledge into clear written communication, especially when emotions become involved. Breeders and owners often undersell good horses, oversell horses emotionally, forget important details, assume buyers just understand, struggle describing temperament accurately, or freeze completely staring at blank screens.

How do you begin? Instead of trying to magically produce polished wording from nowhere, just provide AI with realistic details and ask it to organise the information professionally. Immediately, you have a starting point. Not a final draft, but momentum.

> *Help me write an honest and grounded sales advertisement for a 15hh Gypsy Cob 5 year old gelding broken to harness. Quiet and kind to handle, green, best suited to an experienced person wanting to develop for show or resale. Include temperament, handling manners, and suitability. Avoid hype or exaggerated claims.*

Describe temperament properly

This is one of the most valuable ways AI can help. Many advertisements use meaningless phrases like bombproof, perfect, push-button, child safe, or 'no dirt'. These phrases may create dangerous assumptions because different people interpret them differently. One of the best things a seller can do is explain who a horse suits rather than who might buy it; not who could manage it under ideal circumstances.

> *Help me describe this horse's temperament accurately and professionally for a sales advert. Quiet to handle, sensitive under saddle, gains confidence with routine, not suited to nervous riders.*
>
> *Help me explain suitable rider level for this horse clearly and honestly without sounding negative.*

This becomes especially helpful for green horses, sensitive horses, young horses, competition prospects, family horses, stallions, and horses requiring confident handling. Clear suitability language reduces failed homes dramatically.

Remove emotional wording

Horse people are emotionally attached to their horses. That is normal. But emotional attachment often creates advertisements filled with vague sentimental language or anthropomorphism, instead of practical information.

For example: *Sad sale, available only to the perfect forever home.*

The reality is that all horses need suitable homes, responsible owners, and realistic expectations. Emotional language sometimes clouds practical communication.

> *Rewrite this advertisement to sound warm and inviting while focusing more on suitability and practical information rather than emotional wording.*

That subtle shift improves the integrity of the advertisement dramatically, while still sounding warm and human.

Make headings that stop the scroll

People scroll quickly online. You want a heading that can command attention. AI is extremely good for brainstorming headings and introductory lines that are likely to stand out. Strong headings without exaggeration create curiosity.

Give me 10 attention-grabbing and professional heading ideas for a family-safe Clydesdale gelding, broken to both harness and saddle, which has done heritage displays before huge crowds.

Create sales advertisement headings for [insert text] focused on temperament, education, and suitability rather than hype.

AI can help with Facebook advertising too

Social media advertising requires slightly different wording than traditional sales ads. Facebook readers tend to engage more with conversational language and storytelling while still wanting practical information quickly.

Write a calm Facebook advertisement introducing a young Warmblood mare available to experienced performance home. Ask me what information I need to provide you with. Avoid hype and unrealistic promises.

In a classified advertisement, you might see: 'Stunning mare. Regretful sale. Will only go to best home.' True perhaps. Useful in practice? Not particularly.

After thoughtful AI refinement, it might become: *Show hunter quality 3 year old 15.1h mare, straightforward to handle with good ground manners and sensible temperament. Green under saddle due to age, currently progressing steadily with consistent work. Best suited to experienced competition home wanting a genuine young horse to continue developing patiently.*

Now readers understand something useful. That changes everything.

Advanced prompt:

Theme: Writing honestly about a difficult horse

Suited to: private sellers, dealers, studs, re-homing operations

Act as a professional equestrian copywriter who specialises in honest, ethical horse advertising. I need to write a sales advertisement for a horse that has some known limitations or complications, and I want to find the right buyer rather than make the fastest sale.

Here are the horse's details: [describe breed, age, sex, height, colour, breeding if relevant]. Here is the full, honest picture including any known issues: [describe health history, behavioural traits, past incidents, limitations on use, specialist requirements. Be completely honest here].

Write an advertisement that does the following: leads with genuine strengths and what this horse offers the right person; describes the limitations honestly and specifically, without euphemism; identifies clearly who this horse is and is not suited to; uses a heading that will attract the right buyer and filter out the wrong one; ends with a clear call to action that invites serious enquiries only. The tone should be direct, warm, and completely trustworthy.

Do not use language that softens, hides, or re-frames the known issues. A horse in the right hands is the goal.

For sale: Genuine trail horse

If you are looking for a horse that has actually done the miles, this is him.

Midnight is a 16hh black 15 year old Friesian cross gelding with the kind of experience that cannot be manufactured. He has covered mountain trails, beach rides, bush country, and group rides, and he does it all with the same calm, sensible attitude every time. He camps overnight without fuss, stands quietly on the picket line, and settles like he has been doing it his whole life — because he has.

He is sensible to handle, good in company or alone, and has the solid hooves and bone that serious trail riders appreciate. He is not fragile, and he does not need perfect conditions.

He can be difficult to catch in the paddock. If that is a deal-breaker, he is probably not your horse. If you are experienced enough that it is just 'ho-hum'; you will appreciate everything else he brings.

He would suit an older experienced rider, a true stockman, or a capable family wanting a horse they can trust in the real world.

Horses like this are genuinely hard to find. Priced accordingly at [price]. Located [location]. Serious enquiries welcome [contact details].

Before and after example

Advanced prompt:

Help me write a sales post for a reliable, sound, 15 year old trail gelding. Tone should feel friendly, honest, and professional while clearly explaining suitability, experience and training level. Include the following:

Breed, age, height and colour.

Any specific trail experience: mountain trails, beach, bush, group rides, alone,

Temperament highlights: what makes him sound, easy or trustworthy?

Any quirks or limitations to be honest about?

Ideal buyer: experienced adult, older rider, beginner, family?

Price, location, contact details]

When you have your basics, you can shorten or expand it at will, and include a photo or two, turn it into a flyer or a shorter Facebook advert. Here it is now, shortened from 220 words to 105, just by asking AI to "tighten" it.

For sale: Genuine trail horse

Midnight is a 16hh black, 15 year old Friesian cross gelding with experience that can't be manufactured: mountain trails, beach, bush and group rides, always calm and sensible. He camps overnight without fuss and stands quietly on the picket line and doesn't need perfect conditions. Sensible to handle, good alone or in company, with the solid hooves and bone serious trail riders look for.

He can be hard to catch in the paddock: not a deal-breaker if you appreciate everything else he brings.

Suits an older experienced rider, a genuine stockman, or a capable family wanting a horse they can trust. Hard to find. Priced accordingly at [price], Located [location]. Serious enquiries: [contact details].

Everyday photos and videos of daily happenings around your property keep viewers interested, and develop trust over time. Janice de Gennaro photo.

Chapter 11

AI for studs and farms

(No, not that AI)

Are you carrying far more information in your head than you should? Which mare foals early? Which bloodline matures slowly? Which mare stresses during travel? Which horses hold weight poorly through winter? Which buyers need extra support after purchase? Which foals lacked confidence initially but improved dramatically with consistency? Which clients need follow up? What if an enquirer seeks information about a horse you bred 20 years ago?

Good breeders notice patterns constantly. The problem is that much of this information never becomes properly organised.

It lives in notebooks, phone notes, feed room whiteboards, old diaries, screenshots, memory, and occasionally on scraps of paper floating around the stable office like emotional confetti.

Breeders often function through sheer determination and memory alone. Many become remarkably skilled at managing organised chaos over time. But relying entirely on memory creates problems, especially as breeding programs grow.

Good organisation improves horse outcomes

This is where artificial intelligence becomes quietly but powerfully useful. Organised information leads to better communication, better consistency, better planning, better buyer support, better horse management, and better long term breeding decisions.

It also reduces stress for humans. Overwhelmed breeders often struggle not from lack of knowledge, but from information overload.

I will paste my broodmare notes below. Please summarise recurring temperament patterns, developmental observations, and management trends in her foals, including maturity rates, trainability, confidence levels, and feeding requirements.

Mare and foal records become easier to structure

Breeders collect huge amounts of information over time: covering dates, foaling dates, temperament observations, growth patterns, handling milestones, health and reproductive concerns, veterinary history and competition results,.

The challenge is keeping it accessible and useful. This creates clearer long-term records without requiring hours of manual formatting.

Please organise these foal development notes into a timeline format including confidence, handling progress, feeding changes, and behavioural observations and file under their dam's names as headings. Create a checklist for future foals.

Stable management systems become easier too

Horse businesses run on routines. Feeding schedules, medication routines, farrier schedules, turnout rotations, and exercise programs all matter. Horses thrive on predictability. All of this can be added to routine checklists. Simple systems can reduce mistakes which improves welfare enormously.

Help me create a daily checklist for a small breeding property managing broodmares, foals, and spelling horses. Include a basic feeding and medication tracking template for an orphan foal at various ages until weaning.

Buyer information becomes easier to manage

You may find yourself repeatedly explaining the same information to buyers. Instead of rewriting this information constantly, structure reusable educational resources.

Help me create a buyer information sheet explaining what to expect, and transition management for young horses moving to new homes. Include a simple first week checklist for buyers.

Organise veterinary and health records

AI can assist you to structure vaccination schedules, worming records, injury timelines, feeding changes, veterinary summaries, medication tracking, and recovery observations. (This does not replace veterinary advice.)

Please summarise these veterinary notes of my injured horse into a timeline of treatment, recovery progress, and ongoing management requirements.

Contracts and paperwork

Horse people dislike administrative work intensely, because paperwork feels mentally tiring after physical work, horse care, and daily responsibilities.

Use AI to draft and organise sale agreements, care instructions, welcome packs, boarding policies, client communication templates, contracts, and important documents. This saves enormous mental energy. Some can be uploaded to your website.

> *Help me outline a horse sale agreement covering payment terms, collection and travel arrangements, and buyer responsibilities.*
>
> *Help me create a Bill of Sale template*
>
> *Create a friendly welcome pack outline for agistment clients including stable rules, feeding routines, emergency and contact information.*

Ask AI to compare features

AI can also help you think strategically before investing time and money into systems that may not suit your business. Instead of endlessly Googling random software while drowning in sponsored advertisements and conflicting reviews, you can ask AI specific questions about what you need.

> *Recommend horse management software suitable for a small boarding farm specializing in broodmares, and foaling down services. Include feeding and rugging schedules, veterinary records, owner information, foaling down, stallion servicing and artificial insemination records.]*
>
> *What horse record-keeping programs are best for breeders wanting pedigree tracking, foaling records, and health management?*
>
> *Suggest useful horse management apps suitable for non-technical horse farms wanting easy feeding, medication, and stable records.*

Asking what horse software is best is far too broad. Best for what? Breeding? Competition? Agistment? Training? Veterinary management? Client bookings? Pedigree analysis? The clearer your needs become, the more useful the recommendation usually becomes.

The smartest horse businesses will often combine good, specialised software, thoughtful AI support, strong existing horse knowledge, and clear communication systems together. That balance is usually where the real magic happens.

HorseRecords

Don't try to reinvent the wheel

There is no need to use AI when a good horse specific programme already exists. One of the biggest mistakes people make when discovering AI is trying to force it to provide systems that already work extremely well. Not everything needs reinventing.

There are already some excellent horse-specific record-keeping and management programs available. Systems designed specifically for breeders, horse businesses, competition records, veterinary tracking, stable management, and pedigree organisation.

For example, programs like *HorseRecords* already provide structured horse management systems without breeders needing to build everything manually themselves.

Sometimes the smartest solution is not AI itself but using AI to help you find the right tools more efficiently. That is important because AI should reduce workload, not create more of it.

If a specialised programme already solves the problem properly, use it. Then use AI to support communication, summaries, educational material, organisation, or marketing around that system. That is often the smartest combination.

Programmes like HorseRecords can provide you with excellent means to record daily and seasonal tasks, horse business, breeding and stud information, for hobby horse owners right through to large operations.

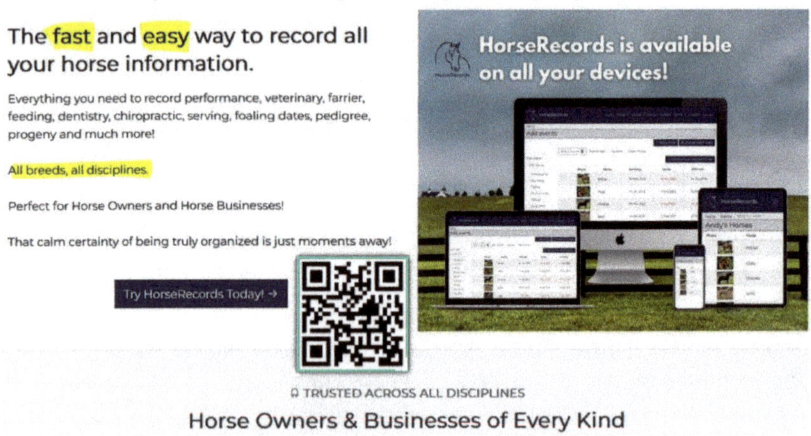

https://www.horserecords.info/

Emotional benefits

Horse people already operate under enormous pressure: physical work, financial pressure, emotional responsibility, animal welfare concerns, and time limitations. When information becomes easier to organise, mental stress reduces dramatically.

Humans generally make better decisions. Relaxed humans benefit horses..

The goal is not handing your entire business over to technology. The goal is improving and reducing unnecessary administrative stress. Good systems still require human oversight, human judgement, and human ethics. The breeders who benefit most are those already trying to improve consistency thoughtfully.

Start with the areas already causing frustration

Do not begin by trying to completely digitise your entire programme overnight. That path leads directly toward overwhelm and emotional collapse somewhere beside the printer.

Focus on achieving small steps. Small successes will increase your motivation.

The next chapter moves on to communication and promotion, specifically, how AI helps breeders present their horses, their stallions, and their breeding philosophy to the wider world.

Scenes around the farm always capture interest

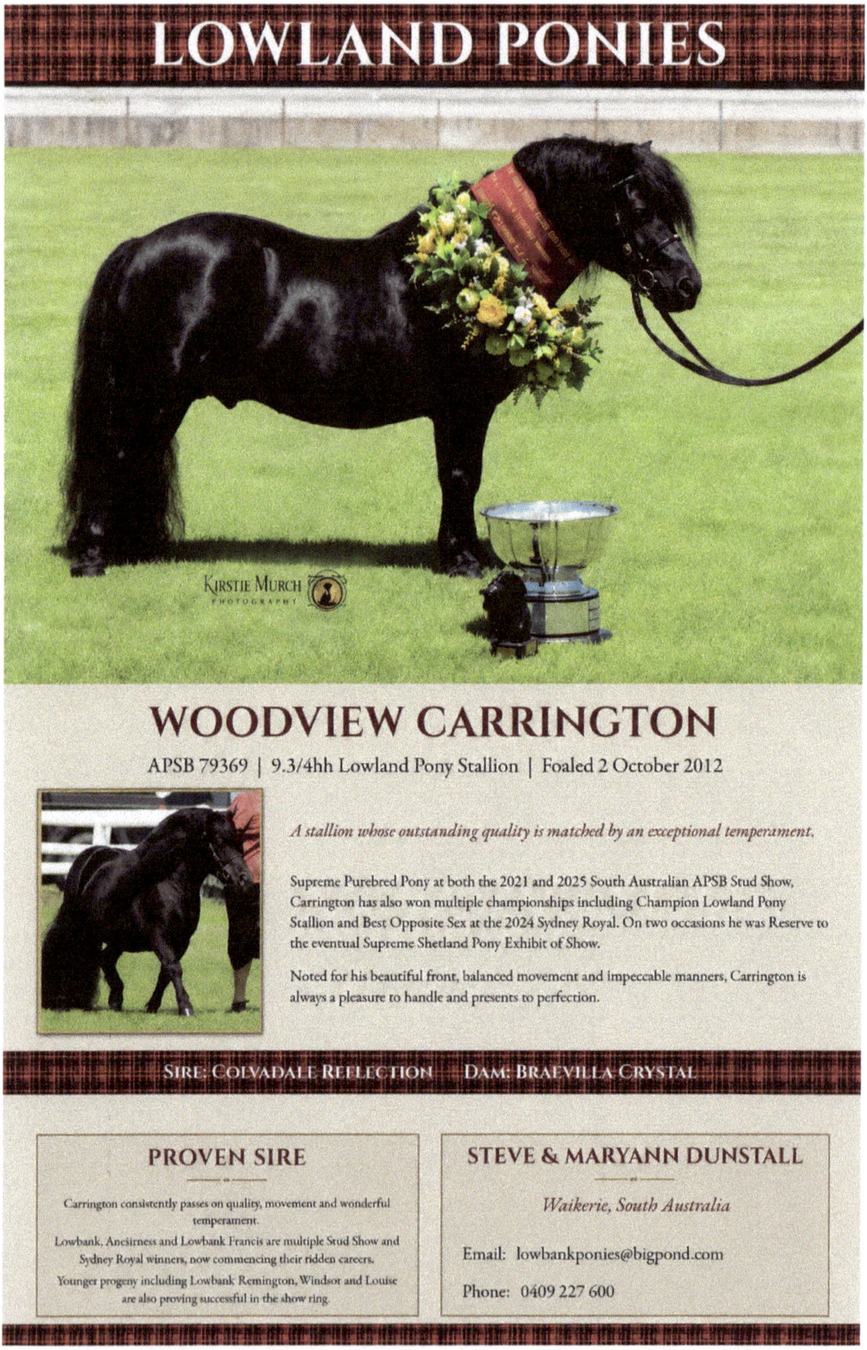

Stud pony advertisement done in Claude Design. The advert isnt perfect, but it shows you what you can achieve in 60 minutes or so. Photo credits Kirstie Murch and Vicki Franklin.

Chapter 12

Stud and stallion promotion

What mare owners actually want to know

Stallion advertising is one of the most fascinating areas of the horse industry because it reveals two things very quickly:

How breeders see their horses. And how well they communicate. Unfortunately, many stallion advertisements drift heavily toward fantasy rather than useful information.

Every stallion is apparently elite, exceptional, rare, world class, a once in a lifetime opportunity!

Meanwhile mare owners are quietly trying to work out basic details like:

- What is his conformation like?
- What is his temperament like? Is he trainable, sensible?
- What are his strengths?
- What has he done under saddle?
- What sort of mares does he suit?
- Is live service or chilled / frozen semen available?
- Has he been tested for genetic diseases?
- What type of offspring does he consistently produce?
- Do they suit amateurs, families or professionals?

As a stud owner, these are basic questions which you might have to repeat to enquirers over and again.

Good stallion advertising is really about communication not just about attracting attention. The strongest stallion advertising usually creates trust through realistic presentation, because mare owners are making important long term decisions.

And stronger breeding decisions shape the future quality of horses over generations.

Why many stallion advertisements fail

A stallion advertisement may struggle in one of two ways. It is either too vague or completely overloaded with dramatic language and cluttered graphics until nobody can absorb the actual information. Sometimes advertisements become so visually chaotic they resemble nightclub posters. Huge fonts. Flames. Sparkles. Logos. Three slogans. Twenty achievements. Tiny unreadable contact details hidden somewhere near the bottom like a scavenger hunt prize. We've all seen them.

Good advertising does not need noise.

Mare owners would also like to evaluate conformation, from a *real* photo, especially as now very few actually visit to see the horse in the flesh.

Good prompting can help breeders structure stallion profiles, stud descriptions, service information, breeding philosophy, temperament summaries, promotional flyers, website wording, Facebook advertisements, and educational breeding content.

> *Help me write a serious stallion profile for our website for a polo stallion standing at stud. Emphasize natural ability, trainability, maternal lines, speed bloodlines, tournaments and awards over x number of years. Include service information and avoid exaggerated wording.*
>
> *Use this information [insert] to construct a stallion A4 flyer to connect with mare owners and send to specialized magazines.*
>
> *Create wording overlays from these images. [insert] Include information on stud fee, location, and bloodlines, along with whether AI, natural service, or frozen semen is available Australia-wide. Highlight multi-champion progeny, quiet and trainable offspring, traditional polo bloodlines, and performance and family versatility.*

One advert can become many formats so repurposing information consistently is going to attract the most attention. Breeders can repeat the same information constantly across different platforms; for example, one well written stallion profile can become a Facebook advert, a one page flyer, a website bio, a magazine layout concept, a social media series, and a mare owner information pack.

Your vision

any breeders have strong philosophies but struggle expressing them clearly. Let AI articulate:

Why you breed, what traits you prioritise, what makes your approach different, why temperament matters, and what type of horses you aim to produce.

Help me write a short statement explaining our breeding philosophy focused on temperament, trainability, and versatile family horses.

Write a professional introduction for our Quarter Horse stallion programme focused on elite reining horses, emphasizing bloodlines, performance awards of the stallions and their progeny, and the types of mares each crosses best with.

This creates emotional clarity alongside practical information. And clients respond strongly to that.

Stallion introduction advert: Help me write a calm and professional Facebook advert introducing our Shetland stallion standing at stud. Focus on temperament, traditional bloodlines, progeny, and family suitability.

Service availability advert: Write a Facebook advert in a professional tone, explaining that frozen semen is available Australia-wide and Internationally, for our Australian Stock Horse stallion [name, contact details.] Include clear collection and service information, breeding enquiries welcome, and contract details.

Write an promotional Facebook post explaining in a conversational tone, why colour will never be the primary focus in our breeding decisions.

Help me create wording for a one-page A4 stallion flyer which emphasizes progeny achievements of our senior barrel-racing stallion.

Suggest a professional magazine advertisement layout for a stallion standing at stud. Include sections for pedigree, achievements, breeding philosophy, service options, and contact information. Then create it as an editable document for me to download.

Good promotion is not just about marketing, it is about responsibility. Good promotion helps mare owners make informed decisions, reduces misunderstandings, and protects reputations, because breeding decisions create long term consequences. A fashionable advertisement may attract attention temporarily, but thoughtful communication builds supporters over decades.

AI should support professionalism, not create fantasy. That means real photos, honest wording, accurate achievements, clear service details, realistic expectations, and consistent branding. The horse world could benefit enormously from less theatre. Good horses rarely need exaggerated marketing.

Foal development updates, mare performance history, stable routines, young horse handling, buyer education, float training, veterinary routines, paddock management, and confidence-building exercises, are all things your followers enjoy seeing.

Instead of staring at your phone every morning trying to think of something to post, you can sit down once and create a month of post ideas, captions, educational themes, seasonal campaigns, and sales advertisement templates in a single session.

Keep the focus on education rather than hype.

> *Generate 30 educational social media post ideas for a horse breeding business focused on producing calm trainable horses for amateur riders.*

Build a sales funnel

That term tends to make horse people uncomfortable. It sounds manipulative. But what it really means is organised communication. A young horse buyer might discover your Facebook page, watch your foal updates, read your educational posts, learn about your breeding philosophy, follow your young horses for months, share some of your posts, and gradually build enough trust to make an enquiry. That is a relationship, not a trick.

If you sell several horses each year, you can create a proper launch system. Instead of throwing up a quick advertisement the moment a horse is ready, you can prepare pre-sale educational posts, progress updates, video content, handling demonstrations, transition information, and buyer FAQs well in advance. Better informed buyers usually make for smoother sales. That alone is worth the effort.

Caption libraries are another thing worth building over time. You probably say the same things regularly; temperament matters, young horses need time, consistency is everything, because those messages are true and your buyers need to hear them.

Give yourself a bank of ready-made content you can use across Facebook, Instagram, newsletters, brochures, and your website, whenever you need it.

> *Create 20 educational Facebook captions focused on intentional young horse development and responsible breeding practices.*

You know your horses. You know your programme and the services you offer. AI helps you say that more consistently, without having to do it all from memory while standing in a muddy paddock with one bar of reception and a horse trying to eat your phone. That balance is usually where the real magic happens.

Advanced prompt:

Theme: FAQ for Sport Horse Website

Suited to: Studs, broodmare owners, stallion stations.

Use this material to build a series of questions and answers for a downloadable FAQ fact sheet for a Sport Horse website. Provide it with an embedded logo, in easy-to-read language targeting persons wanting to know more about Warmblood breeding.

1. What do these abbreviations and studbook codes mean? [paste/describe pedigree].

2. How do I register a foal with [studbook name], and what's the difference between provisional and full papers? What is the range of costs involved?

3. What's the difference between a dam line and a sire line, and why does it matter for breeding decisions? Why are mares considered the most important side of a pedigree?

4. Explain the different levels of grading systems a horse might fall under. What is a linear score?

5. What does this mare's performance record (dressage/jumping scores) tell me about her as a broodmare? [paste record].

6. What is 'nicking' in Warmblood breeding, and please give some examples. How do I research it for my mare's specific line?

7. How do I read a stallion's breeding values or index scores, and what do the different figures represent? How likely is it that assessors agree within small margins on scores given? How much value should I place on these scores?

8. Based on this mare's conformation and pedigree, what traits should I look for in a complementary stallion [Ask me relevant questions including temperament, height.] that may have commercial sale value? Ask me questions based on breeding values and index scores and any other relevant statistics.

9. I want to compare Warmblood statistics between [Country A] and [Country B] Given that studbooks like KWPN, Oldenburg, and Hanoverian use different scoring scales, inspection criteria, and indexing systems, how do I normalise raw numbers so the comparison is meaningful rather than misleading?

10. Ask me questions to evaluate costs for breeding a foal, based on (Country / Location) from conception to sale at auction, at a premier sale, based on a colt considered suitable for passing its 3-year-old grading. What price would I need to achieve to break even, and is this a realistic scenario, based on recent prices?

11. Name which websites I should go to to find additional information?

12. What are excellent books and videos on this subject?

Advanced prompt:

Theme: Complete season promotion package from one session

Suited to: stud operations, stallion owners

Act as an equestrian marketing consultant with expertise in thoroughbred, warmblood, and pony breed promotion. I want to build a complete promotional package for my stallion's upcoming breeding season in a single session.

Stallion details: [name, breed, age, height, colour, sire, dam sire, performance record or progeny achievements, service fee, location, contact]. His particular strengths as a sire are [describe — e.g. temperament, movement, bone, fertility, progeny success in specific disciplines]. His ideal mare type is [describe]. His stud fee is [insert] and the season runs [insert dates].

Please produce the following as a complete package:
1. A website stallion profile of 200-250 words in a warm, professional tone that speaks directly to mare owners making a serious breeding decision.

2. A Facebook and Instagram announcement post of 80-100 words with a strong opening line.

3. A mare owner FAQ covering the five questions I most commonly receive before booking.

4. A short email of 150 words to send to past clients and known contacts announcing the season. All four pieces should be in consistent voice and tone, and all should reflect genuine, specific qualities of this horse, not generic stallion advertising language.

5. Mare owner information pack to send to mare owners once a booking is made.

Chapter 13

Organisation without becoming corporate

Reduce communication overload and misunderstandings

Most horse businesses do not fail because people lack horse knowledge. They struggle because administration quietly consumes enormous amounts of time and energy. Horse businesses are often running two completely different operations simultaneously.

One physical. One administrative. The physical side usually receives most of the attention because horses obviously require immediate care. The administrative side quietly piles up in the background until people suddenly find themselves overwhelmed, exhausted, and replying to client messages at midnight with approximately the emotional stability of an overtired broodmare in fly season.

Messages, bookings, cancellations, invoices, client questions, lesson scheduling, buyer enquiries, follow ups, reminder emails, entries and registrations, have a way of accumulating faster than the work itself.

And all of this usually happens while somebody is feeding horses, holding a lead rope, driving a float, walking through mud, carrying hay, or searching for paperwork that definitely existed yesterday.

Most horse businesses still operate manually, particularly smaller breeding farms and training businesses. Many systems still rely on memory, handwritten diaries, text messages, screenshots, verbal arrangements, and the occasional "I thought you booked that."

While this sometimes works on a small scale, it becomes increasingly stressful as you grow. Disorganisation creates missed communication, scheduling confusion, delayed responses, client frustration, and administrative burnout.

You repeatedly answer the same questions, such as what riders should bring, what happens during lessons, what feeding routines are required, what cancellation policies apply, what agistment includes, and what experience level is suitable.

Create a fun welcome email for new agistment clients explaining feeding routines, stable expectations, safety procedures, farm rules and communication guidelines.

Write a calm and non-formal cancellation policy for riding lessons and clinics that maintains clear boundaries while remaining approachable.

On-boarding

Many of your clients will arrive with different expectations, different levels of horse knowledge, and different assumptions about how things work. On-boarding (orientation or induction) means giving your new clients the information they need before confusion has a chance to develop. A good on-boarding system reduces stress for everyone by answering questions before they are even asked.

AI can help you create welcome packs, stable guidelines, buyer preparation documents, farm rules, lesson information sheets, new client FAQs, and horse care instructions, all written once and used repeatedly as handouts or welcome emails.

I run a horse riding and trail business offering weekend camps for adult groups of no more than twelve riders. I need a complete new client welcome pack that covers communication expectations, appointment procedures, horse handling rules, and payment policies.

The pack should include a registration and information email, a waiver, a simple client guide covering what to pack and bring, expectations of behaviour and safety, a course map outline and general camp orientation, information on camp leaders, and weekend activities. There can be a bonus useful knots section.

The tone should be warm, practical, and confidence-building for riders who may be new to this environment. Please ask me step-by step questions as we go to help refine each section to suit my specific service before you write it. Organise some sections into booklet form, others can be standalone.

Advanced prompt:

Theme: Full client communication suite in one session

Suited to: trainers, riding schools, agistment operators, horse professionals with ongoing client relationships

Act as a business communication consultant with experience in service-based businesses and the equestrian industry. I want to build a complete suite of reusable client communication templates in a single session, so I never have to write these from scratch again.

My business is [describe e.g. a dressage training yard / an agistment property / an equine massage practice]. My typical client is [describe horse and owner]. My tone of voice is [e.g. professional and warm / direct and practical / calm and educational].

Please write the following six templates, all in consistent voice:

1. Initial enquiry response: acknowledges their interest, sets expectations, and invites them to book a call or visit, and why to book a call [insert qualifications and experience].

2. Booking confirmation: confirms details, tells them what to bring or prepare, and sets the tone for the relationship.

3. Pre-visit or pre-lesson preparation note: practical instructions and what to expect, disclaimer and waiver

4. Post-visit follow-up: thanks them, summarises key points from the session, and includes a clear next step.

5. Payment reminder : firm, professional, and non-apologetic.

6. End-of-relationship or departure note which closes the relationship warmly and professionally regardless of the circumstances. Bracket wherever I need to insert specific names, dates, or details.

Lesson planning

Riding instructors carry an enormous amount of knowledge about how people learn, how horses respond, and how to build confidence progressively. The challenge is that much of that knowledge lives in their heads rather than in any organised system, which makes it difficult to maintain consistency across students, hand over to relief instructors, or scale beyond a certain number of lessons per week.

You can describe your teaching approach, the level of your student, and what the lesson needs to achieve, and ask Claude or ChatGPT to help you build a structured lesson plan, a progressive term program, or a set of exercises suited to a particular rider and horse combination.

It is a starting point that saves you an hour of planning time and gives you something to adapt rather than something to invent from scratch.

Create a four-week riding lesson progression plan for beginner riders focusing on confidence, balance, horse handling, and consistency. Give me a solid starting point that keeps my students progressing rather than repeating the same lesson indefinitely.

Create a six-week progressive lesson plan for a nervous adult beginner rider on a calm school horse. Focus on building confidence, position, and basic control before introducing trot work. Include brief session goals and one key exercise per lesson.

Create a simple student progress tracking template for a riding school. Include columns for student name, current level, horse allocated, skills achieved this term, skills to work on, and instructor notes. Keep it practical and easy to update after each lesson.

Write a friendly end-of-term progress letter for a riding school student aged eight who has progressed from walk work to confident trot and is ready to begin canter preparation next term. The tone should be warm, encouraging, and specific. Turn this into a template I can use in the future.

Help me create a complete administrative system for a two-day riding clinic for twelve participants. Include a registration information email, a pre-clinic preparation note for participants, a daily schedule template, a disclaimer and waiver form outline, and a post-clinic feedback survey. The clinic focuses on confidence-building for amateur riders.

Advanced prompt:

Theme: Full riding school communication and planning suite

Suited to: riding instructors, coaches, equestrian centres, and anyone teaching riders at any level

Act as an experienced equestrian business consultant with specific knowledge of riding school administration and student communication. I want to build a complete set of reusable resources for my riding school so that lesson planning, student tracking, parent communication, and clinic organisation all have consistent, professional systems behind them.

My riding school is [describe, for example: a small private yard with ten students / a commercial riding school with fifty students across multiple levels / a specialist dressage coaching business]. My students are [describe age range and experience level]. My teaching style and tone are [describe, for example: calm and methodical / encouraging and progressive / technically focused].

Please build the following resources one at a time, waiting for my feedback after each before continuing:

1. *A six-week progressive lesson plan template for a beginner student that I can adapt for any horse and rider combination.*

2. *A waiver for parents to sign, and rules of behaviour on site*

3. *A student progress tracking template suitable for a full term of lessons.*

4. *A parent communication pack including a term welcome letter, a mid-term update template, and an end-of-term progress letter.*

5. *A clinic registration and administration pack including a participant information email, a pre-clinic preparation note, a daily schedule template, and a post-clinic feedback survey.*

Keep all wording consistent in tone throughout, and leave clear brackets wherever I need to insert specific names, dates, or details.

Advanced knowledge of AI doesn't necessarily make you a better horseperson. Technology can only do so much, without the input of fellow horselovers to make it real. Use it to work smarter, not harder.

Part III

Work smarter, not harder

The following chapters are more advanced applications

Chapter 14

AI on your phone

Some basics

Most of this book has assumed you are sitting at a desk with a laptop or computer in front of you. The reality for many horse people is quite different. You are in the paddock. You are in the feed shed. You are sitting in the cab of a truck waiting for a horse to load. You are rugging at dusk with one hand and trying to reply to a student enquiry with the other.

The phone is how most horse people now run their business, and the good news is that AI works just as well from a phone as it does from a desktop. In some ways it works better, because the features built into modern smartphones add capabilities that a desktop does not have.

Your phone already has AI built in

Before downloading anything new, it is worth knowing that your phone already includes AI tools you may not have thought of as AI at all.

Voice to text is available on every modern iPhone and Android phone. On an iPhone, tap the microphone icon on the keyboard and speak. On Android, do the same. The phone transcribes what you say into typed text in real time. This means you can dictate a client email, a Facebook post, a set of mare notes, or a prompt to Claude while walking between paddocks, and have it typed out by the time you reach the gate.

This one feature alone is worth stopping to practice. Many horse people find dictating far faster than typing, particularly with muddy or gloved hands. The accuracy of modern voice to text is realistic, though it is worth reading through before sending, as the occasional word will be misheard.

Siri on iPhone and Google Assistant on Android can also be used to set quick reminders, make calls, send messages, quick look-ups and answer basic questions hands-free, while you are working with horses.

Lists

If you are like me, you make daily to-do lists, weekly or monthly checklists, shopping lists, and appointment lists.

This habit worth building is simple. The moment you think of something, open Notes on your phone, and add it. The value is getting it out of your head before the next horse demands your attention.

The Notes app

The Notes app costs nothing, needs no login, and is already sitting on your phone waiting to be useful. You may have already been using it.

To-do lists:

- Open the Notes app on your phone.
- Tap the pencil icon to start a new note.
- Type a title at the top, something like "Monday jobs" or "Before the farrier comes."
- Tap the checklist icon (it looks like a small tick or bullet point) to turn your note into a tappable list.
- Type each task on a separate line.

After you complete each task, tap the circle beside it to check it off. Anything unchecked rolls over as a visual reminder every time you open the note.

Shopping lists:

- Open Notes and start a new note.
- Title it "Feed store" or "Rural supplies" or whatever suits you.
- Turn on the checklist format as above.
- Add items as you think of them, not just when you are about to leave.
- At the store, check each item off as it goes in the trolley.

Uncheck everything when you get home so the list is ready for next time.

Speak your prompt

Both Claude and ChatGPT have free apps available for iPhone and Android. They work identically to the desktop versions, and both support voice input, which means you can speak your prompt rather than type it.

To use voice input in the Claude or ChatGPT app, tap the microphone icon in the message field, speak your prompt clearly, and the app will transcribe it and send it as your message. This makes it practical to ask AI for help while you are working, as long as you have a moment to listen to or read the response.

Some uses work particularly well from a phone while you are working with horses:

- Drafting a quick reply to a client enquiry without going inside to find a computer.
- Generating a social media caption for a photo you have just taken and want to post.
- Dictating a vet update.
- Asking for help with wording on something while the thought is fresh.
- Setting up a reminder for a follow-up call or vaccination date.

> Thanks to my son-in-law, Andy, for this excellent tip for desktop users:
>
> Windows has a built-in shortcut that can transform how you interact with AI. Pressing Windows + H activates voice dictation system-wide, meaning you can speak your prompt directly into any AI chat window without typing a word.
>
> You can read more about this approach here:
>
> https://djangoandy.com/2026/03/18/speak-first-prompt-later-using-windows-h-to-supercharge-your-ai-prompts/

Your phone camera as an AI tool

The camera on a modern smartphone is more than a camera. Several AI-powered features are built directly into it or available through free apps.

Google Lens, available on Android and also as an app for iPhone, allows you to point your camera at something and ask AI to identify or explain it. For horse people this has practical applications. You can photograph a plant in a paddock and ask whether it is toxic to horses. You can photograph a product label and ask for an explanation. You can photograph a document and have the text extracted and converted into something you can edit or copy.

Apple's built-in Live Text feature works similarly on iPhone, allowing you to photograph any text, including handwritten notes, and have it converted to editable text automatically.

Neither of these tools is perfect, and anything involving horse health or pasture safety should be verified with your veterinarian rather than relied upon from an AI identification alone. But as a first step toward identifying an unknown plant or understanding an unfamiliar product, they are informative starting points.

Manage horse photographs on your phone

One of the most practical AI features available on modern phones is photo organisation. Both iPhone and Android can search your photo library by content, which means you can search for the word 'foal' and the phone will show you photographs it has identified as containing foals, or search for a horse's name if you have been consistent about adding it to photo descriptions.

For horse people who accumulate large numbers of photographs, this makes finding the right image for an advertisement, a social media post, or a sales enquiry significantly faster.

When photographing horses on your phone for promotional use, a few good habits make a significant difference. Film in good natural light, ideally in the morning or late afternoon rather than harsh midday sun. Keep the background as uncluttered as possible. Film horizontally for website and email use, vertically for Instagram reels, and either, depending on usage, for Facebook.

Once you have your photo, use the crop feature to centre or enlarge the subject, and other editing features to remove background, messy objects, turn to black and white, or sepia for effect.

Canva on your phone

Canva has a free mobile app for both iPhone and Android that works well for quick edits, posting to social media, and creating simple graphics on the go. Most of the desktop features are available in the app, and for horse people who manage their promotional material primarily from a phone, it is worth downloading and exploring.

Using AI from a phone while managing horses requires the same judgement you would apply to any task that splits your attention. Drafting a message while a horse is tied safely is sensible. Dictating notes while leading a horse is not. The phone is a tool, and like every tool around horses it deserves to be used at the right moment and put away at the wrong one.

Screen capture

On a phone, pressing the side button and volume button together on an iPhone, or the equivalent on Android, captures the full screen, though most phones now also allow you to scroll and capture a longer page in one image. You might find it easier to just take a photo.

Advanced prompt:

Theme: Setting up AI tools on your phone for horse business use

Suited to: horse people who manage their business primarily from a smartphone

Act as a practical digital assistant helping a horse person set up AI tools on their phone to use in their day-to-day work. I manage my horse business largely from my phone and I want to use AI more effectively while I am working, without needing to sit at a desk to do it.

Please ask me the following questions one at a time and wait for my answer before continuing:

1. *What phone do you use, iPhone or Android, and what is the approximate age of the device?*
2. *What tasks do you currently spend the most time on when managing your horse business, for example communication, record keeping, social media, or scheduling?*
3. *Which of these tasks do you most often try to manage while you are physically working with horses or away from a desk?*
4. *Do you currently use voice to text on your phone, and if so, how comfortable are you with it?*
5. *What AI apps, if any, do you already have installed?*

Once I have answered all questions, please suggest a phone setup plan that includes which AI apps to download, which built-in phone features to activate, and two or three specific ways I could use AI from my phone this week to save time on the tasks I described. Keep the suggestions practical and suited to someone working outdoors with horses.

Short question: Do you have an email signature on your phone? This is one of the most important features you can have for contact availability. If you are not sure how to do this, ask AI how to set it up and what to put on it. Seriously, go do it!

This one is suitable for advertising riding instruction, a catalogue for tack, or a background for a brochure.

I gave AI a prompt to use the following elements: mounting block, riding boots, gloves, and helmet. This is what it gave me.

I gave the same instructions to my photographer, and this is what she gave me. One is literal, perfect; the other imperfect, real. Two results, with two completely different interpretations and intent behind them.

One has never been near a horse.

The paddock office: saddled up, ready to ride, and still a prisoner to the phone. What might you use a photo like this for? Photo credit: Janice de Gennaro.

Chapter 15

Create repeatable systems

Build systems that work

Using AI occasionally is one thing. The real shift happens when you start building systems around it.

Most horse businesses operate reactively. You write the sales ad when you need it, answer buyer emails when you finally sit down, and repeat the same information from memory, over and over.

That works on a small scale, but becomes time consuming as your programme grows, particularly when you are managing multiple mares, young stock, waiting lists, buyer communication, veterinary schedules, and client education all at once.

The question stops being "How do I use AI?" and becomes "How do I build systems that reduce stress and improve consistency?" That is a very different question, and a much more useful one.

A system is nothing complicated. It is a repeatable way of doing something, a foal handling routine, a student on-boarding process, a worming schedule, a sale preparation checklist. Good systems reduce decision fatigue, particularly during breeding and show season.

AI is excellent at creating standard operating procedures, which really just means "how we do things here." You already have these routines. AI helps get them out of your head and onto paper.

Create a standard operating training procedure for introducing young horses to trail obstacles, including safety considerations, confidence building, timing, and gradual progression.

Help me create a stable procedure document for managing agisted broodmares and foals during severe weather conditions.

These can become incredibly important for training, consistency, emergency preparation, and client education.

You are probably answering the same owner questions repeatedly: current feed, handling experience, rugging, transition advice. A reusable communication template answers all of it once, improving buyer confidence, reducing miscommunication, and cutting the emotional exhaustion of constant repetition. Try something like:

Create a reusable information template to an owner of a horse under rehabilitation, including feeding routines, handling notes, exercise levels, veterinary history, rugging preferences, and transition advice back to its owner's location.

Beyond communication, AI can help you build a genuine knowledge library. Without documentation, it disappears when you step away. The horse industry loses extraordinary amounts of practical knowledge this way.

Help me organise my breeding notes into categories including temperament patterns, maturity rates, handling sensitivity, feeding management, and bloodline observations.

For larger farms and busy equestrian facilities, workflow systems add another layer: foaling preparation timelines, sale horse preparation, veterinary scheduling, staff induction, emergency procedures. These help you move from reactive to organised, particularly when family members, staff, casual workers, vets, and trainers are all involved.

Create a six week workflow checklist preparing young horses for professional sale advertisements, including grooming, photographs, video preparation, handling consistency, and buyer education material.

Create a daily communication template for staff managing broodmares, foals, feeding routines, medications, and paddock changes.

Good systems reduce unnecessary pressure. But the goal is never to become so focused on technology that your warmth and personality disappear. Your horse business should still feel human, personal, experienced, and thoughtful.

You deserve better than "Ask Karen, she remembers everything."

***

Advanced prompt:

Theme: Bushfire emergency plan for horse properties

Suited to: horse property owners, stud managers, agistment operators, rural horse keepers

Act as an experienced rural emergency management consultant with specific knowledge of horse property management in bushfire-prone regions of Australia. You are going to help me create a comprehensive, property-specific bushfire emergency plan for my horses.

Before building the template, ask me the following questions one at a time and wait for my answer before proceeding to the next:

1. *How many horses do you have, and what are their individual needs; age, health, handling difficulty, and any that require special consideration such as foals, mares in foal, or horses with mobility issues?*
2. *What emergency contacts do you have; vet, farrier, neighbours, local rural fire service, and do you have a designated person to call if you cannot be reached? What happens if you are at work?*
3. *What water is available on your property; tanks, dams, troughs, and what is your backup plan if mains or pump power fails during a fire event?*
4. *Where are your safe yards or holding areas; how far from the main stable, are they accessible by float or truck, and are they clear of overhanging trees and fencing that could become a hazard?*
5. *Where are your safe evacuation centres? Show a map of quickest route.*
6. *What transportation do you have available; how many horses can you move in one trip, do you have a backup float or truck arrangement, and how long does it take to load your horses under stress?*
7. *How many people can you reliably call on during an emergency, and what are their roles? Who handles horses, who drives, who manages gates and equipment?*
8. *What is your trigger point for beginning evacuation; fire warning level, kilometres distance, visible smoke, and have you identified your evacuation route including an alternative if the main road is cut?*

Once I have answered all questions, compile my answers into a clear, structured emergency template I can print, laminate, and keep in my float and feed shed. Include a one-page quick-reference checklist at the end with actions listed in priority order.

Advanced Prompt:

Theme: Standard operating procedure for one complete process

Suited to: a family farm needing someone to look after the place in an emergency / long-awaited holiday for professional yards, training barns, studs with staff or locums.

Act as a business operations consultant with experience in service businesses, equestrian operations, and staff management documentation. I need to document one complete operational process from my business as a formal Standard Operating Procedure, something clear enough that a competent person with horse experience but no knowledge of my specific operation could follow it correctly.

The process I want to document is [choose one and describe it in detail: e.g. horse intake and assessment when a new horse arrives / client on-boarding for a new livery / competition preparation from two weeks out to day of travel / foaling watch procedure / stallion collection and dispatch / emergency procedures].

Here is how we currently do it, as I would explain it to someone new: [describe your current process in plain language, including any steps that are informal or undocumented].

Please produce:

1. A formal SOP with numbered steps, each written in clear imperative language.

2. Decision points clearly flagged: where a staff member must make a judgement call and what criteria to use.

3. A checklist version of the same procedure suitable for printing and ticking off in real time.

4. A list of anything that is currently undocumented that I should add before this SOP is complete. Format the document so it could be handed to a family member, locum or new staff member on their first day.

Booking and follow-up calls without the back-and-forth

If you offer lessons, consultations, or training reviews, the constant messaging back and forth to find a time that suits everyone can eat up far more of your day than it should. Calendly solves this neatly. You set up your available times once, connect it to the calendar you already use, and share one simple link on your Facebook page, website, or in your email signature. Clients pick a slot that works for them, and it lands straight into both calendars. No more "are you free Tuesday?" messages while you're holding a hoof pick.

You can set up different booking types depending on what you offer. A lesson might be sixty minutes in person, a phone consultation twenty minutes, and a video review call thirty minutes over Zoom. The free version of Calendly covers most small operations comfortably. If you offer a wider range of services with different durations and prices, Acuity Scheduling is worth a look as well.

For remote lessons, training discussions, or follow up calls, Zoom lets you sit down face to face with a client wherever they are. The part that makes a real difference for horse people is the option to record the call.

Recording matters because clients often need to revisit what was discussed. They might want to re-watch a conversation about a horse's behaviour, a training plan, or feeding advice without having to ask you to explain it all again.

If a client sends through video of their horse and you talk through it together on the call, that recording becomes something they can keep and return to. It cuts down on the follow up messages asking you to repeat yourself, and it gives clients a genuine sense that they are getting real value from the session.

To record, click "record" once the call begins, choose whether to save it to the cloud or to your own computer, and Zoom will process it once the call ends. You can then send the link through afterwards, and the client does not need a Zoom account to watch it.

It is good practice to mention at the start of the call that you are recording it, out of courtesy.

When you connect Calendly and Zoom together, the whole process becomes automatic. A client books a video review slot, Calendly creates the Zoom meeting and sends the link to both of you, and after the call you can send through the recording as a simple follow up.

That follow up often becomes a nice extra touch that clients remember, and it builds the kind of trust that keeps people coming back.

Online forms: contracts and applications

Every horse business generates paperwork. Sale contracts, agistment agreements, lesson waivers, breed society registration forms, stallion service contracts, boarding applications. Traditionally this means printing, signing, scanning, emailing, and chasing people for the version they forgot to sign. Multiply that across every client, every season, every foal registration, and the hours add up quickly.

Online forms solve this. A client fills in their details on a form on your website or through a shared link, signs electronically, and the completed document lands in your inbox ready to file. No printing, no scanning, no "I'll bring it next time."

Who this benefits

This is not just about saving yourself effort. A properly filled-in form, typed rather than handwritten, benefits everyone downstream.

Breed societies receive registration applications that are legible and complete, rather than guessing at someone's handwriting or chasing missing fields. Veterinary practices receive consistent information. Your own records become searchable rather than living in a folder of scanned PDFs with names like "scan0047.p df".

Tools to create online forms

Google Forms: free, simple, and good enough for most purposes, waivers, enquiry forms, simple applications. It does not handle signatures well, but for information gathering it is genuinely all most people need.

JotForm: offers more templates suited to contracts and includes basic e-signature functionality on its free plan. It also has horse industry-specific templates already available, including boarding agreements and liability waivers.

AI can help you turn an existing paper form or contract into a clear digital version, suggest fields you may have missed, and draft the wording for sections such as terms and conditions or liability clauses (*which should still be checked against your local legal requirements*).

Another benefit is that you can mark fields as required, meaning the form cannot be submitted until every section is completed. No more contracts arriving with the emergency contact left blank, or a sale agreement missing the buyer's address because they ran out of time and meant to fill it in later. The form does that checking for you, before it ever reaches your inbox.

> *I have a paper-based agistment contract. Here is the text: [paste your existing contract]. Please identify all the fields that need to be filled in, suggest a logical order for an online form, and flag any sections that may need updating for clarity.*

Getting started

Pick one form you need repeatedly, a sale contract, a lesson waiver, a boarding application, and convert just that one first. Once it is working and people are using it without confusion, move on to the next. This is the same small-steps approach the book has used throughout: one form done well, is more practical than five half-finished ones.

Advanced prompt:

Theme: Setting up and embedding an online form

Suited to: anyone wanting to add a form to their website for the first time

Act as a patient, practical assistant helping someone with no technical experience set up an online form for their horse business.

Please ask me the following questions one at a time and wait for my answer before continuing:

What is this form for, for example a sale contract, a lesson waiver, an enquiry form, or a boarding application?

What information do I need to collect, and which fields absolutely must be filled in before the form can be submitted?

What platform is my website built on, for example WordPress, Wix, Squarespace, or something else, or do I not have a website at all?

How comfortable am I with editing my website directly?

Once I have answered all questions, give me clear, step by step instructions for creating the form, marking the required fields, and either embedding it on my website or sharing it as a link, whichever suits my situation best.

Quick tip: Sometimes you get a Gobbledygook error message which means nothing (unless you're a techno wizz), like this:

INFO(CSS-007): /OEBPS/style.css(873,3): Font-face reference "https://2407c8e8-5f35-44c2-8e3d-24351f68e6ce.epubcheck.w3c.org/OEBPS/fonts/RobotoCondensed-Regular.ttf" refers to non-standard font type "application/x-font-ttf"

I just ask Claude to fix the error, and re-upload it. Easy!

What you bring to AI is far more valuable than what AI brings to you.

Chapter 16

Email sequences and automation

Stay connected without the constant effort

If you have ever tried to stay consistently in touch with clients, buyers, students, or followers, you already know how quickly it becomes exhausting. You write something. People respond. Life with horses gets busy. Weeks pass. And before long the silence has stretched out longer than you intended, and the relationship quietly cools.

This is not a failure of effort. It is a reality of the horse world, where the horses themselves will always come first, as they should.

What email sequences and automation offer is a way to stay connected with the people who matter to you, without requiring you to sit down and write something fresh every single time. You build the sequence once, in your own voice, and it continues working for you in the background while you get on with the horses.

This chapter is not about becoming a slick internet marketer or turning your horse business into something it is not. It is about using a simple, practical system to communicate more consistently, building trust over time, letting people find their own way to your products or services without feeling pushed.

What is an email sequence?

An email sequence is a series of emails written in advance and sent automatically to people who have signed up to hear from you. They might have downloaded a free resource, joined your mailing list at an event, enquired about a horse, enrolled in a clinic, or simply expressed interest in what you do.

Rather than you remembering to follow up manually, the sequence does it for you. Each email goes out at a set interval, in the order you decide, and it keeps the relationship moving forward without you needing to think about it each time.

The sequence can be used for almost any purpose. Selling a product or digital download, welcoming new students to a course, educating buyers before they breed to a stallion, staying in touch with a mailing list, or sharing knowledge gradually with people who want to learn from you.

Building your email list

Before any sequence can run, you need a list of people to send it to. This sounds obvious, but it is the step many horse people have not yet taken in any organised way.

You probably have more contacts than you realise. Past buyers, current clients, event attendees, people who have enquired about horses or services, followers who have reached out on social media, students from workshops or clinics, tradesmen you deal with, hay suppliers and the like. These people already have a relationship with you. They are not yet in one organised place.

The most common ways horse businesses grow their lists are through a simple sign-up form on their website, a free resource offered in exchange for an email address such as a buyer guide or foaling checklist, a link in a Facebook post or bio directing people to subscribe, a sign-up sheet at events or clinics, or asking existing clients directly whether they would like to receive your updates.

Whatever method you use, always ask for a first name alongside the email address. It costs the person nothing and allows every email you send to open with their name, which makes a genuine difference to how personal it feels.

The most important rule around list building is *consent*. Only add people who have actively agreed to receive emails from you. Adding someone without their knowledge is both ineffective and in most places legally problematic. Always give them a means to unsubscribe.

A smaller list of sincerely interested people will always outperform a larger list of people who did not ask to be there.

What is a CSV file?

If you already have contact details stored somewhere, whether that is in Gmail, a spreadsheet, an old database, or another email platform, you will likely need to move them into your email automation tool as a CSV file.

CSV stands for comma-separated values. It sounds technical but it is just a plain text file that stores information in rows and columns, like a very basic spreadsheet. Each row is one contact. Each column is a piece of information about that contact, typically their first name, last name, and email address. The file ends in .csv rather than .xlsx or .docx.

Most programs that store contact information can export a CSV file. In Gmail, for example, you can go to Google Contacts, select the contacts you want to export, choose the option to export them, and save the file as a Google CSV or Outlook CSV format. This gives you a file you can then upload into an email platform.

This assumes that you have added all your names into Google Contacts over time. If not, start right away.

If your contacts are currently sitting in a spreadsheet, you can open the spreadsheet in Excel, Numbers, or Google Sheets and save or export it as a CSV file. Make sure at a minimum you have one column labelled email and one labelled first name before you export. Those are the two fields most platforms require.

Advanced prompt:

Theme: Preparing a CSV file from existing contacts

Suited to: anyone moving existing contacts into an email platform for the first time

Using Claude: Act as a patient and practical digital assistant helping a non-technical person prepare their contact list for import into an email marketing platform. I have existing contacts stored in [describe where, for example: Gmail, a spreadsheet, my phone, another email platform] and I want to import them into Kit as a CSV file.

Please walk me through the exact steps to export my contacts from where they currently are, prepare the CSV file correctly with the right columns and headings, and then import it into Kit. Assume I am not technical and explain each step as though I have not done this before. If anything could go wrong, tell me what to watch for and how to fix it.

Getting started in Kit

Kit, formerly known as ConvertKit, is one of the most widely used email platforms among educators, authors, and small business owners, and it is the platform this chapter references throughout. It is well suited to horse people because it is straightforward to use, built around sequences and automations from the ground up, and has a free plan that is genuinely helpful.

As of 2026, Kit's free newsletter plan supports up to 10,000 subscribers at no cost and includes unlimited emails, landing pages, and forms. This is one of the most generous free tiers available in email marketing and is more than enough to get started and run a full sequence.

When your needs grow or you want access to more advanced automation features, at the time of writing, paid plans begin at around $33 per month for up to 1,000 subscribers, billed annually.

To get started, go to kit.com and create a free account. You will be asked for your name, email address, and a little information about what you plan to use the platform for.

Once inside, the main areas you will work with are Subscribers, where your contacts live; Sequences, where you build your email series; Automations, where you connect the trigger that starts a sequence to the sequence itself; and Forms or Landing Pages, which are the sign-up points where new people join your list.

Importing your contacts into Kit

Once you have your CSV file ready, importing it into Kit is straightforward. Go to the Subscribers section of your Kit account and click the Add Subscribers button. Choose Import CSV from the options that appear. You can drag your file directly into the window or click to select it from your computer.

Kit will then ask you to map your columns, which means telling it which column in your file corresponds to which field in Kit. Match the email column to email address and the name column to first name. Columns you do not need can be left unmapped and will be ignored.

Before completing the import, Kit will ask you to add a tag to this group of contacts. A tag is a label that helps you keep track of where people came from. Something like *Imported from Gmail* or *Existing clients* works well. This makes it easy to find this group again later if you need to.

Kit also handles duplicate email addresses automatically, merging them rather than creating duplicate records, so there is no need to manually clean your list before importing, if you are unsure whether some addresses appear more than once.

Make it sound like you

People stay subscribed to those they enjoy reading. They refer friends to people who feel genuine. And none of that happens if your emails sound like they were written by a committee or generated by a machine with no understanding of horses.

Your voice, your experience, and your perspective are the reasons people chose to follow you in the first place. A sequence that loses that will lose people too.

Before you use any prompt to build a sequence, make sure you have entered a set of instructions that teaches Claude how you write, what words you use, what tone you carry, and what you care about.

The chapter on sounding like yourself covers this in detail. With that in place, the emails Claude helps you write will feel like you sat down and wrote them yourself on a good day, not like a generic marketing template with your name pasted in.

The structure that works

A sequence of fifteen emails is a practical and manageable number. It covers enough ground to build real trust, handle questions before they become doubts, and give people genuine reasons to take the next step, without overwhelming anyone or burning through your ideas in the first week.

The first five emails build the relationship

These are not selling emails. Their job is to make people glad they signed up. You share something useful, something honest, something that reflects your experience and your perspective on horses. You are not asking for anything. You are just showing up and being worth reading.

Most people send one welcome email and then go quiet. Five emails of genuine value puts you well ahead of that, and by the time you reach the middle of the sequence readers already feel they know something about how you think.

The next five emails introduce your offer

These emails do the work of explaining what you have to offer and why it might matter to the reader. But they do it through stories, examples, and honest explanation, not through pressure or manufactured urgency.

Each of these emails follows the same simple shape. It opens with a real problem the reader is likely experiencing. It backs that up with something from your own experience or with a specific example that shows you understand the territory.

It then points toward a way through, showing what becomes possible when the problem is solved. And it closes with one clear, single next step. Not three options. Not a wall of links. One invitation.

The goal of these emails is not to convince people who are not interested. It is to make it easy for the people who already are interested to take action. There is a meaningful difference between those two things, and understanding it changes how you write.

The final five emails follow up with people who nearly said yes

This is the part most people skip entirely, and it is often where the most value sits.

Some people will open your emails, click through to your product or booking page, and then close the browser without doing anything. Life interrupted them. They got distracted. They meant to come back and forgot. These are warm, genuinely interested people who did not complete the step.

A short series of follow-up emails sent only to people who clicked but did not buy or book, gives them a gentle second chance. No pressure. No guilt. Just a quiet reminder that the door is still open. Done well, this feels considerate rather than pushy, and it consistently reaches people who were already most of the way there.

Personalising your emails

Most email platforms allow you to address each reader by their first name automatically, pulling it from the information they provided when they signed up. This is a small thing that makes a genuine difference. An email that opens with a person's name feels like a letter, not a broadcast.

To use this in Kit, you type the merge tag for first name into your email wherever you want the name to appear. Kit uses the format {{ subscriber.first_name }} and will replace it with each person's actual name when the email is sent. If a subscriber's first name is missing, you can set a fallback word such as "there" or "friend" so the email still reads naturally.

What this is not

Automation does not mean disconnection. The risk with any automated system is that it becomes an excuse to stop engaging personally altogether, and that would be a loss for a horse service where relationships matter so much.

Think of the sequence as the foundation, not the ceiling. It keeps communication moving in the background. But the real conversations, the replies, the personal responses, the moments when someone shares something about their horse and you write back properly, still happen. The sequence just means you are never starting from complete silence.

And before you send any sequence to your list, read back through every email and ask yourself one honest question. Would this make me want to buy my own product, join my own course, or reach out to my own business? If the answer is yes, you are ready.

If something feels flat or unlike you, go back and ask Claude to adjust it. The sequence should feel like the best version of a conversation you would have with someone you met at a horse event; warm, knowledgeable, grounded, and worth their time.

Advanced prompt:

Theme: Building a 15-email sequence in your own voice

Suited to: horse educators, breeders, trainers, course creators, authors, and anyone with a product, service, or knowledge to share

Act as an experienced email sequence strategist who understands how to build genuine trust with an audience before making any offer. I want to create a 15-email automated sequence that sounds exactly like me, uses my real experience and knowledge, and communicates warmly and directly without sales pressure or manufactured urgency.

Please ask me the following questions one at a time and wait for my answer before continuing:

1. What is the product, service, course, or offer this sequence will lead toward? Describe it in plain language including what it does, who it is for, and what problem it solves.

2. Who is your audience? Describe the person who will be reading these emails, including their experience level, their frustrations, and what they are hoping to achieve.

3. What is the single most common mistake or misunderstanding your audience has about this topic? This will shape several of the trust-building emails.

4. What proof, results, or real experience do you have that shows your offer works? Include specific examples, outcomes, or stories if you have them.

5. What objections do people most commonly have before they buy or sign up? List the main ones honestly.

6. What is the one clear action you want readers to take at the end of the sequence? Be specific, for example: click to purchase, book a call, join a waitlist, or download a resource.

7. Upload two or three examples of your own writing so Claude can match your voice throughout the sequence.

Once I have answered all questions, please do the following:

First, show me the full 15-email plan before writing anything. List each email by number with a working title and one sentence describing its purpose. Wait for my approval before writing a single email.

Then write each email individually, one at a time. Each email should open with a real problem the reader is likely experiencing. Back it with something from my own experience or a specific example. Show the way through, and close with one clear, single next step. Address the reader by first name using the placeholder [First name] at the start of each email.

The first five emails should build trust and deliver genuine value with no selling. The next five should introduce the offer authentically and handle objections through stories and real examples. The final five should be shorter follow-up emails for people who showed interest but did not yet take action. No hype, no urgency. Just a genuine conversation I would have with friends.

Every careful step counts: tracking training progress means measuring the small wins as much as the big ones. This image was captured off a video, at precisely the right moment for use on Facebook. It came out with remarkable clarity.

Chapter 17

Numbers and spreadsheets

More powerful than most people realise

Most horse people use spreadsheets occasionally, for end of year financials, foaling dates, competition entries, or the occasional breeding record. They open one, type in some numbers, and close it again.

AI can transform a basic spreadsheet into a powerful analysis tool, without requiring any technical knowledge, coding experience, or understanding of complex formulas. This chapter is about knowing enough to ask AI to do the work for you.

A disorganised list leads to duplicate messages, missed follow-ups, and no way to track what is working.

> *Please set up a proper tracking spreadsheet for [insert from the following or describe your own: pasture and property management / horse purchase contacts and follow-up / breeding records and mare cycles / competition and show entry tracking / expert interviews and sources / social media content planner for horse topics performing well / equine therapy and rehabilitation progress tracking / horse transport and agistment contacts / stable and yard maintenance schedules and contractors / emergency contacts list].*
>
> *Please include all relevant columns and explain how to use it effectively.*

Spreadsheets are already part of horse life

Whether you use Excel, Google Sheets, or Numbers on a Mac, you are already working with spreadsheet software. The challenge is that most people use maybe five percent of what these tools can do.

But now, instead of learning the formula yourself, you describe what you want and ask AI to write it for you. Then you paste it in. That is the whole process.

What AI can build in a spreadsheet for you

The list is longer than most people expect.

- Foaling date calculators from covering date
- Feeding cost trackers that update automatically
- Mare cycle tracking with predicted return dates
- Competition schedule planners with countdown timers
- Horse entries events schedule with rider names
- Sales inquiry logs with follow-up reminders
- Veterinary expense summaries by horse or year
- Worming and vaccination schedule trackers
- Pedigree databases with search functions
- Coefficient of Inbreeding calculators from pedigree data

Every one of those can be built by describing what you want and asking AI to write the formulas. You do not need to understand the formula. You need to paste it into the correct cell and press Enter.

Not all AI platforms handle numbers and structured data equally well.

For analysing data you upload, ChatGPT with its Advanced Data Analysis feature (also called Code Interpreter) is currently the strongest option. It can run calculations on your spreadsheet rather than just reading it, which means it produces verified results rather than estimates.

For reading images, pedigree screenshots, handwritten records, photos of old stud books, Claude and GPT-4o both handle image content very well. Upload the image, describe what you need, and ask for the analysis.

For large datasets such as multiple years of stud records, population-level COI analysis across a breeding program, the most reliable approach is to export your data as a CSV file and upload it with a specific analysis request. The more structured your data, the more useful the analysis.

How to ask AI to build a spreadsheet formula

The key is being specific about what you have and what you want. Vague requests produce vague results.

> *I have a Google Sheet with horse names in column A, covering dates in column B, and I want column C to automatically calculate the expected foaling date based on an eleven-month, 11 days gestation period. Please write the formula for column C.*

That gives AI everything it needs. The platform, the column layout, the specific calculation required. The result will be a formula you can paste directly into your spreadsheet without modification. Specificity is important here.

Using data you already have

You probably have years of data sitting in notebooks, old spreadsheets, or stud management software. It may be year by year, of by individual horses. Foal weights, temperament observations, growth rates, performance results, and veterinary notes, all of it collected over time and almost none of it formally analysed.

AI can help you look at that data and find patterns you may never have noticed by eye. For example, you could paste several years into an AI conversation and ask:

> *I am going to paste my foaling records for the past six years. Please look for patterns in birth weight/height, growth rate, and final weight/height by sire line. Identify any trends worth noting.*

> *This spreadsheet contains my feeding costs by horse for the past three years. Please identify which horses cost the most to maintain, which months are most expensive, and whether there are any patterns worth investigating.*

These are not complicated requests. They are asking AI to do what it does well; read information, organise it, and look for connections.

A starting point

If spreadsheets feel unfamiliar, start with something simple and immediately useful rather than jumping straight into calculations.

A good first project might be a basic mare record spreadsheet with columns for mare name, last covering date, expected foaling date calculated automatically, foal born date, foal description, and weaning details. Ask AI to build the formula for the foaling date column, and the spreadsheet becomes a useful working tool immediately. (Note: many programmes, including HorseRecords have inbuilt foaling calculators).

From there, the same spreadsheet can grow. Add columns for veterinary costs, temperament observations, growth notes, or sale price. Each new column is just another question to ask AI.

The goal is not to become a data analyst. The goal is to stop losing useful information to notebooks, memory, and scraps of paper floating around the feed shed.

Remember: because some programmes can do this for you, such as HorseRecords, you could instead download and paste its performance data for a particular horse, and ask AI to summarise the achievements into a format for your website or a sales advertisement.

Using this performance data from HorseRecords, attached, please summarise the horse's [name] achievements into a format [template/summary/article] for my website.

Using this source 4-gen pedigree template, please create a 5-gen fillable template, using the same colour and format, widening the page so the fifth generation has room instead of becoming ant writing.

Pedigree analysis and the coefficient of inbreeding

This is one of the more specialised but valuable uses of AI for breeders. Done properly, a coefficient of inbreeding calculation tells you something a glance at a pedigree never can: not just which ancestors repeat, but how much genetic material those repetitions are likely to concentrate in the foal. The maths has always existed. What AI changes is how accessible it has become.

Using Wright's Path Coefficient method, the calculated COI for Chalani Star Wars is approximately 3.87 percent. That is a mild figure, sitting comfortably in the low range, and indicates a well-planned cross with reasonable genetic diversity. The primary contributors are Rannock, who appears seven times across the pedigree (some in the 5th generation, not shown here), and Master Herbert, who creates two valid loops. Rannock's own inbreeding coefficient of 12.5 percent, the result of a half-sibling mating, was factored into the calculation using the (1 + FA) multiplier built into Wright's formula.

One ancestor that might appear relevant at first glance is Tintaras Callisto. After careful checking, however, Tintaras Callisto does not form a valid sire-to-dam loop for Chalani Star Wars and was correctly excluded from the calculation. This is a common point of confusion, and it matters.

Why "count the names" does not work

Most people, when they first look at a pedigree, try to count how many times a name appears. That is not how inbreeding is measured.

I do serious pedigree analysis in my classic work, *The Thinking Horse Breeder*. See information at the end of this book.

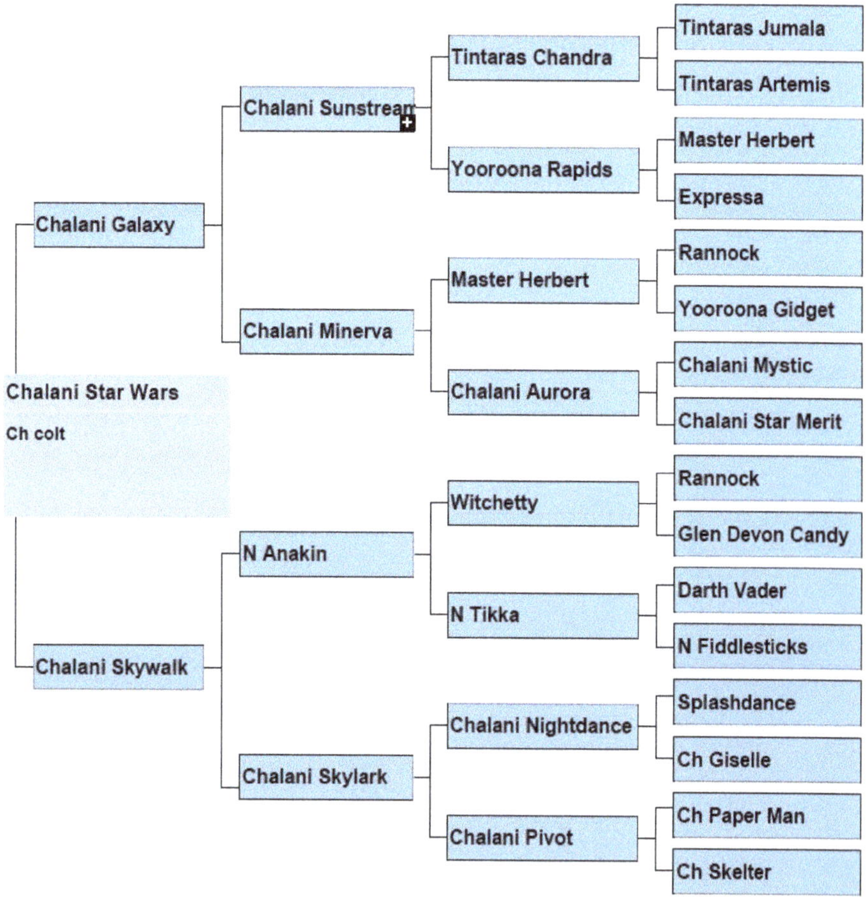

Using this example: Chalani Star Wars:

Wright's method counts complete journeys: from the sire, back through the pedigree to a common ancestor, and then forward again through the dam's side to the foal. Each of those return trips is called a loop, and each loop contributes a separate figure to the final COI. A horse whose name appears four times may contribute several loops. A horse that appears twice may contribute only one, or none at all, if both appearances are on the *same side* of the pedigree.

There is also a rule that catches people out regularly. Once you have identified the nearest common ancestor in a path, you stop there. You do not then add the ancestors sitting behind that horse as additional loops, unless they also appear elsewhere in the pedigree through a completely independent route. In this pedigree, Tintaras Chandra and Expressa appear only because Chalani Sunstream appears on both sides. They are Chalani Sunstream's ancestry, not independent contributors. Counting them separately would inflate the COI and produce a misleading result.

Then I asked AI to look at the pedigree of this proposed foal:

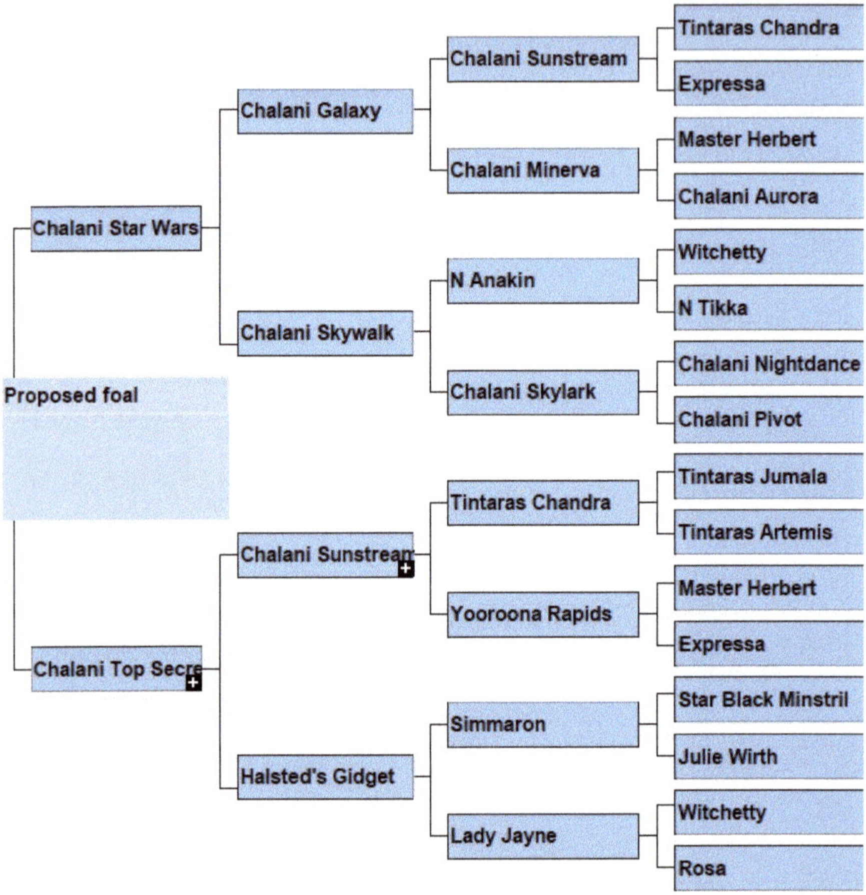

The formula:

Wright's Path Coefficient formula is:

F(x) = Σ (0.5)^(n1 + n2 + 1) × (1 + FA)

Where n1 is the number of steps from the sire back to the common ancestor, n2 is the number of steps from the dam back to the same ancestor, and FA is the inbreeding coefficient of that ancestor, if known. The diagram below shows how this works for a single loop through Chalani Sunstream.

Chalani Sunstream is the sire of Chalani Top Secret on the dam's side, and the grandsire of Chalani Star Wars on the sire's side through Chalani Galaxy. That gives n1 = 2 and n2 = 1. The exponent becomes 2 + 1 + 1 = 4, and the contribution from that one loop is (0.5)^4 = 0.0625, or 6.25 percent of the proposed foal's COI. Chalani Sunstream is the single largest contributor in the proposed cross, accounting for more than half the total figure on his own.

When the same AI session was used to evaluate the proposed foal the COI came back at 11.96 percent. This is provisional on the pedigree information available as if you looked at 10 generations, the coefficient might be different. For example, many of these horses go back to a horse called Panzer, multiple times.

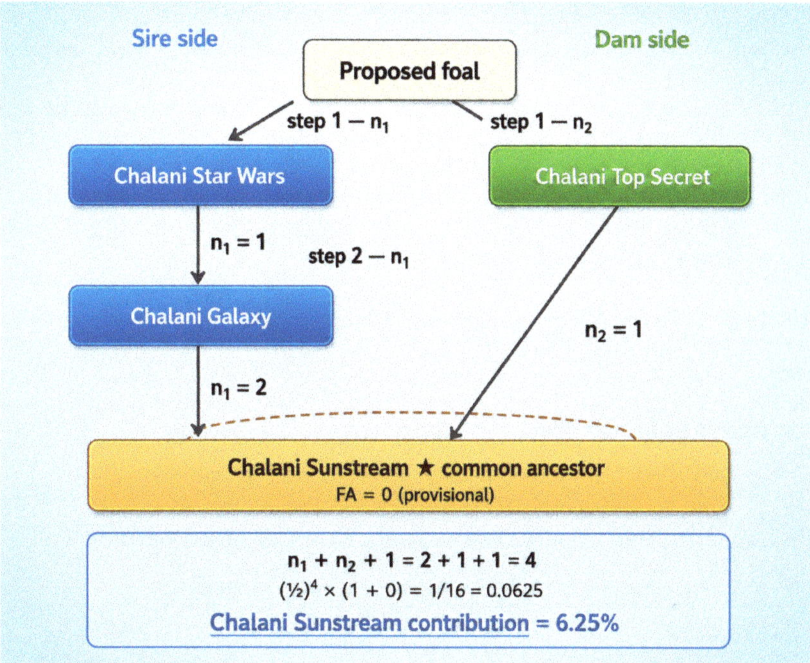

Wright's Path Method: N loops for Chalani Sunstream in the pedigree of proposed foal

Common ancestor – Contribution Percentage of total

0.119629 × 100 = 11.9629 percent, rounded to 11.96 percent.

It is worth noting that Chalani Mystic, who appears in the fifth generation as the sire of Chalani Aurora, is by a grandson of Rannock out of a Rannock mare. Witchetty is also by Rannock. That breeding adds to Rannock's already significant presence across both sides of the pedigree and is part of why his contribution remains substantial even when Chalani Sunstream is the headline figure.

What 11.96 percent actually means:

his is line-breeding in the traditional sense, used deliberately to fix type. As a rough guide to interpreting any COI figure:

- Under 5 percent: mild, common in most breeds
- 5 to 10 percent: moderate line-breeding, deliberate concentration on certain bloodlines
- 10 to 15 percent: significant line-breeding, worth knowing well before you proceed
- Over 20 percent: heavy inbreeding, increased risk of genetic problems

At 11.96 percent, this proposed foal sits in the significant line-breeding band. That is not alarming on its own. What it means is that the horses being concentrated, Chalani Sunstream, Rannock, and Master Herbert in particular, carry real weight in this pedigree. The question is not whether the number is safe. It is whether those horses are worth concentrating on, and whether the *individual animals* involved are robust enough to carry it.

That is a pedigree judgement only an experienced horse person can make. The maths gives you the honest starting point. Your breeding decisions are yours, and yours alone.

Coefficient of Inbreeding – proposed foal

This cross produces a COI of 11.96% – in the low to mid range of moderate inbreeding.
The dominant source is Rannock, contributing 3.96%. The next largest is Chalani Sunstream at 6.25%.
Smaller contributions come from Master Herbert (0.98%) and Witchetty (0.78%).
Tintaras Chandra and Expressa are not listed separately as they are fully accounted for within Chalani Sunstream.
This is a manageable level of inbreeding for an outcross within the Chalani line.

Common ancestor	Contribution	% of total
Chalani Sunstream	0.062500	52.2%
Rannock (F_s = 12.5%)	0.039551	33.1%
Master Herbert	0.009766	8.2%
Witchetty	0.007813	6.5%
Total COI – proposed foal	**0.119629**	**100%**

I asked AI to prepare this as a chart.

What AI does well here, and where to check its work

AI can read a pedigree image, identify every ancestor appearing on both sides, trace the connecting paths, apply the formula, and present the results in plain language. Work that would take a person a considerable time to do by eye across several generations can be done in a single conversation. It can also create the images I used here.

What it cannot do reliably without guidance is know which ancestors to exclude, how to handle an ancestor who only appears through a closer common ancestor, or whether a particular horse's own inbreeding should be factored in. AI is an excellent tool for this work. It is not yet a sufficient one on its own.

Advanced prompt:

Theme: Calculating Wright's COI

Suited to: breeders, stud owners, and anyone researching bloodlines. (Attach your pedigree image before sending).

Using the pedigree in the attached image, calculate the Coefficient of Inbreeding (COI) for [horse name or proposed foal] using Wright's Path Coefficient Formula:

F(x) = Σ [(0.5)^(n1 + n2 + 1) × (1 + FA)] where:

- *n1 = number of steps from the sire to the common ancestor*
- *n2 = number of steps from the dam to the common ancestor*
- *FA = inbreeding coefficient of the common ancestor (assume 0 unless I specify otherwise)*

Instructions:

Read the pedigree carefully and identify every ancestor that appears on both the sire side and the dam side.

For each common ancestor, identify every connecting path. Once you reach the nearest common ancestor in a path, stop there. Do not count ancestors sitting behind that horse as additional loops unless they independently appear elsewhere in the pedigree through a different route.

List each common ancestor and show the calculation for every path through it separately, including the n1 and n2 values and the path contribution.

If any common ancestors are known to be inbred themselves, apply the (1 + FA) multiplier. If I have provided their COI, use it. If not, note that their own inbreeding is unknown and the final figure may be understated.

Sum all path contributions to produce the final COI as both a decimal and a percentage. Present the results clearly: common ancestors identified, number of valid paths through each, individual contributions, and the final total.

Note any ancestors you were uncertain about and ask whether I can provide further information. [Add here: the names of any ancestors whose own inbreeding is known, and their COI if available. For example: Rannock is the result of a half-sibling mating, FA = 0.125.]

Colour genetics calculations

Breeders who work with colour genetics know how quickly the calculations multiply. Five loci, each segregating independently, and suddenly you are staring at 32 possible genotypes before you have even thought about epistasis. AI handles that arithmetic instantly, but more usefully, it explains what the numbers mean in the paddock; what colours to expect, in what proportions, and why.

Advanced prompt:

Theme: equine coat colour prediction from known, partial, or untested genotypes

Suited to: breeders working with colour genetics, stud managers planning crosses, breed society record-keepers, and anyone using genetic test results or pedigree observation to predict foal colours

Act as an equine geneticist with expertise in coat colour inheritance. Before calculating, ask me the following questions one at a time and wait for my answer:

- *What is the known or tested genotype of each horse, if available?*
- *If the full genotype is unknown, what are the visible colours of each horse's sire, dam, and any progeny already produced?*
- *Has either horse had genetic colour testing, and if so, which loci were tested and what were the results?*

Once you have enough information to establish the most likely genotype for each horse, state your assumptions clearly and flag any loci that remain uncertain. Then:

- *identify all loci involved and state the ratio at each locus,*
- *apply 2^n to calculate all possible genotype combinations and their probabilities,*
- *convert those genotypes into expected visible coat colours, accounting for epistatic interactions between loci,*
- *present the final result as both a percentage and a simplified ratio, and*
- *explain in plain language what a breeder should expect to see in the foal crop.*
- *where uncertainty remains around any locus, provide a range of possible outcomes rather than a single answer. My horses are: [insert what you know about each horse].*

These examples are simply here to show you what is possible. Specialist programmes exist for these calculations, but the point is not about genetics. Whatever your situation with horses, you can develop prompts that fit your specific needs. You just need to know what you are trying to solve. AI can even prepare the prompt. Then it does the rest.

Other calculations

Beyond breeding calculations, there are everyday formulas that horse people use constantly, often without realising they are doing maths at all.

The most common is body-weight estimation. Using a girth measurement and a body length measurement, both in inches, (the formula girth^2 x length / 330) gives an estimated weight in pounds. For anyone without access to a weighbridge, this is genuinely useful, particularly when working out feed amounts or medication dosing accurately rather than guessing.

Once you have a weight estimate, feeding becomes much simpler. As a general guide, a horse needs roughly 1.5 to 2 percent of its body-weight in total feed per day, with the majority ideally coming from forage.

> *How much should I be feeding my [insert weight] horse daily, for maintaining condition (currently ideal condition)?*
>
> *Help me understand how much I am feeding [insert your hay in flakes or bales and hard feed etc] and what needs increasing or decreasing.*
>
> *Help me select lower cost substitutes for what I am currently feeding my horses during a drought year. Ask me questions that will help to answer this, including availability, transport logistics, price, weights, protein content and other nutritional factors, supplements needed, horse usage etc.*

Water intake is another area where a rough formula helps. A horse typically drinks between 25 and 50 litres of water per day, depending on size, workload, temperature, and how much moisture is already in their forage. This becomes particularly relevant when planning for transport, isolation yards, or dry seasons.

For anything involving medication dosing, supplements or energy requirements for performance horses, AI can explain the concepts and terminology clearly, but the final numbers should always come from your vet or a qualified equine nutritionist. As with everything else in this book, AI is there to help you understand the numbers, not to replace the professionals who should be calculating them for your individual horse.

For those of you who would like more detailed methodology for using AI on your phone, check out this wonderful resource:

The Equine Advantage

For those of you who would like to learn more about social media video reels for beginners, especially for Instagram and TikTok, I can do no better than to recommend Jay Hunt's wonderful **The Equine Advantage** courses for people using their smartphone.

They are real value for money as she explains every step in detail through video. Additionally, she provides a series of 100 quick, copyright-free reels you can adapt for your own use.

You can find her here: https://hub.equinevideopro.com/

Has this book given you even one useful idea, saved you an hour, or simply made you feel more confident in managing the complexity of horses and everything that comes with them?

I would be extremely grateful if you would take a few minutes to leave a review on Amazon or Goodreads.Reviews matter enormously for independent authors. They help other horse people find the book, and they help me keep writing.

You do not need to write much. A sentence or two about what was useful, what surprised you, or how you plan to use what you have learned is more than enough.

Thank you for reading, and as always, for keeping the horses in mind.

Chapter 18

Using Design platforms

Making your horse business look as good as it is

QR codes

You have almost certainly seen them, those small square patterns of black and white dots appearing on business cards, flyers, posters, advertisements, and social media posts. They are QR codes, and for horse businesses and creators they are one of the simplest and most effective ways to connect your printed or digital material directly to your website, your product pages, your booking forms, or anywhere else you want your audience to land.

Instead of typing a long web address, a QR code does the work instantly. One scan and they are there. Unlike normal printed material, this gives you a live link the audience can go to directly. You can link your audience directly to your products and pages.

Always test before publishing

Before you post, print, or send anything containing a QR code, always test it first. QR codes can lose clarity if resized after creation or compressed during export. A code that looks sharp on screen may scan poorly once it has been through a printing or file compression process.

Book Brush

BookBrush is a design tool built specifically for authors and creators and is really useful for creating promotional material such as mockups, social media graphics, and advertisements. If you have written articles, a book or plan to, BookBrush is worth exploring. It also includes the QR code generator described here. It is more narrowly focused than Canva but does that narrower job very well. I use BookBrush a lot for its ease of use.

Free QR code generators are also available in Canva and Adobe Express, and through dedicated sites such as QR Code Generator. The process is identical: paste your link, download the image, then use it wherever you need it.

How to create a QR code in BookBrush

BookBrush includes a simple built-in QR code generator. If you are already using BookBrush for your promotional material, creating your QR code there keeps everything in one place.

- Log into BookBrush and open the tools menu
- Select the QR code generator
- Paste in the URL you want the code to link to
- Choose your size and download the finished code as an image file
- Copy and paste it onto your promotional material, email signature, or social media graphics as needed

How to check a QR code

You do not need a separate app. On both iPhone and Android, the built-in camera handles this automatically:

- Open your camera app as normal
- Point it at the QR code and hold steady
- A notification bubble will appear showing the link
- Tap it and it will open in your browser

Canva

Throughout this book we have discussed artificial intelligence tools that work primarily through conversation. You type a question and the system responds. Canva is different. Although Canva now includes AI features, it is primarily a design platform like BookBrush.

It can help create brochures, newsletters, educational handouts, social media graphics, event programs, stallion promotions, sale advertisements, presentations, clinic flyers, event schedules, certificates and a wide variety of other resources. In plain terms, Canva helps make information look better.

One of the challenges within the horse world is that people often wear multiple hats. A breeder may also be responsible for marketing. A riding instructor may also create educational resources. An event organiser may also produce programs, schedules, and promotional material. A breed society may need newsletters, membership information, educational resources, and historical publications. Hiring professional designers for every project is often unrealistic.

It is important to be realistic. Canva is easier than professional design software but that does not mean it is instantly easy. Most people require time to become comfortable with the templates, layouts, images, fonts, colours, resizing, file exports, folders, and design organisation. Canva provides thousands of templates. Many can be adapted to horse related purposes with only minor changes. A riding school newsletter can begin with a newsletter template. A clinic flyer can begin with an event template. A breed society publication can begin with a magazine layout.

Imagine a riding school preparing for a school holiday program. The organiser needs promotional graphics, registration forms, information sheets, certificates, and social media posts. Traditionally, this might require several different programs. Canva allows much of it to be created in one place. Canva provides a practical way to achieve that without requiring professional design skills.

Colour and font consistency

Stay true to your branding. One of the most effective things any horse service can do is choose a small set of colours and fonts and use them consistently across everything. Decide on your colours and write down their hex codes, which are the six character codes that identify a precise colour, so you can match colours exactly every time.Save them somewhere easy to find. Three colours is plenty.

The same applies to fonts. Choose one for headings and one for body text, and resist the temptation to try something different each time you open a new design. When your newsletter, your social media graphics, your flyers, and your email signature all look like they belong together, your business immediately appears more established and appealing, even if you are working entirely from your phone and a free account.

Common beginner mistakes

- Using too many fonts.
- Using too many colours.
- Adding too much information.
- Overcrowding pages.
- Trying to make everything look impressive.

Simple designs are usually more effective. Clarity should always come before decoration.

Note: This prompt is used in ChatGPT or Claude, not typed directly into Canva. Use the suggestions AI gives you as your design brief, then open Canva to build the finished product.

Advanced prompt:

Theme: Canva design assistant for horse industry materials

Suited to: anyone creating visual content, for horse owners or professionals

Act as a Canva design assistant for the horse industry. I am creating a [insert your design type here, for example: newsletter, brochure, flyer, information pack, social media graphic, or presentation] for an audience of horse owners and horse professionals.

Please suggest:

1. *Layout ideas suited to this format*
2. *Recommended section headings*
3. *Image and photography style recommendations*
4. *Colour themes appropriate for the horse industry*
5. *Ways to improve readability and visual flow*
6. *Any design improvements that would make this more professional and appealing to my audience*

Exporting your finished design

When you finish a design, you will need to export it as a file before you can use it anywhere. The format you choose matters. If you are unsure, PDF for print and PNG for images will cover most situations correctly.

- **Export as a PDF** when you are sending something to a printer or sharing a document people will read on screen. PDF preserves your layout and fonts reliably across different devices.
- **Export as a PNG** when you need a high quality image, particularly one with a transparent background such as a logo or graphic element. PNG is the better choice for most print-ready images.

- **Export as a JPEG** when file size matters more than perfect quality, such as for social media posts or website images where a smaller file loads faster.

  Quick Tip: You can upload an image to AI, and ask it to change the image size to the size you need. In most cases it will do a good job.

Canva on your phone

Canva has a free mobile app for both iPhone and Android that works very well for quick edits, posting to social media, and creating graphics on the go. Most of the desktop features are available in the app, though working on complex layouts is easier on a larger screen. For horse people who manage their business primarily from a phone, the app is worth downloading.

Adobe Express

A free design platform that works in a similar way to Canva. It is worth knowing if you are already using other Adobe tools such as Photoshop or Lightroom, as everything sits within the same account. It includes templates, image editing, social media graphics, and a QR code generator. For most horse people, Canva and Adobe Express overlap considerably in what they offer. Either works well. The best one is the one you will actually use.

> **For free, legally usable images** to accompany your content, Wikimedia Commons (commons.wikimedia.org) holds millions of photographs, illustrations, and historical images that are free to use, and share, many without requiring attribution. Always check the individual licence on each image before using it, as conditions vary between files.

Working with images without using design tools

The good news? You no longer need to do these in the design tools themselves. You can give AI your ideas, let it give you the plan and the finished product to combine photographs with text, graphics, headings, and layouts. Simple works better than complicated.

Good images are usually clear, well lit, uncluttered, and relevant to the message. You can use them for your 'hero' section, the prominent area at the top of a webpage, landing page, email, or advertisement. Like a shop window, it creates the first impression and should quickly tell visitors what the page is about, who it is for, and what to do next.

A good hero communicates value immediately, guides action, and gives the page a professional, purposeful feel. Without one, visitors can be left wondering what they are looking at. The key principle is simple: the hero is not decoration, it is *direction*. If visitors instantly understand what you offer and where to click next, the hero is doing its job.

ChatGPT and Claude Design can be used to create excellent work *without you needing to learn about design tools.* It is a matter of learning the language, (like, header/footer, hero, render) to more accurately set up your prompts while giving it your own photos. Claude Design allows you to 'edit' as you go along, and set up editable downloads.

Protect your work with a simple watermark

If you regularly share horse photographs, documents, worksheets or educational material online, a discreet watermark is a simple way to protect your work and promote your name or business. It reminds people who created the content and makes it less inviting to copy without permission.

Creating a professional watermark once required software such as Photoshop or Lightroom; now AI can design one in minutes. Once you have a transparent PNG, you can use it in almost any image editor, presentation tool or online design platform, and you only need to create it once.

For PDFs, you can upload the file and ask Claude or ChatGPT to add a watermark to every page, such as "Draft only", "Not for distribution", "Confidential" or "Licensed to: John Smith". You can also add a footer, such as "© John Smith. All rights reserved. Copyright protected material", or "© John Smith. This PDF is licensed to one purchaser and may not be copied, shared or redistributed without permission."

Advanced Prompt:

Theme: Watermarks

Suited to: business owners, amateur photographers, designers, anyone wanting some control over their work

Design a clean, professional watermark for my horse business, featuring [NAME or website or logo] in a style that's elegant, modern and understated. Keep colours to black, white or a single dark tone so it works on both light and dark photographs, with a transparent background, no surrounding box or border, and high enough resolution to stay readable at around 200 to 400 pixels wide.

I need a black, white and a subtle grey version, a horizontal layout for the bottom corner of photographs, and a stacked version for portrait images.

The overall look should be professional, timeless and suited to horse photography, books, educational resources and social media; save it as a transparent PNG for use in my image editing software.

Chapter 19

Video creation and short reels

Show and tell

There is a reason short videos are suddenly everywhere. People watch them. Even horse people who insist they do not really do social media somehow still end up watching a three minute hoof trimming reel at 11:40pm while holding cold tea and waiting upon an overdue mare.

Short videos now drive enormous visibility on Facebook, Instagram, reels, TikTok, and YouTube Shorts. Many horse people are massively under-using them because they assume video creation requires expensive equipment, technical knowledge, professional editing skills, or a teenager living permanently inside Canva.

It does not. AI has changed this dramatically. You can now create simple, useful, professional short videos with your phone, a few photographs, small video clips, basic editing apps, and AI prompts for scripting and structure. And importantly, this can be done without becoming fake, over polished, or emotionally exhausting.

Why short videos work so well

Horse people like seeing reality. Movement, temperament, handling, routine, personality. A short video communicates those things far faster than paragraphs ever can. One calm thirty second clip of a horse loading quietly, a foal being handled, a stallion standing relaxed, or a child grooming safely, tells buyers more than fifteen dramatic paragraphs about exceptional trainability and athletic potential. A quiet clip of a mare with her foal, a horse being rugged, a training moment, or a before and after confidence story, will often perform far better precisely because it feels genuine.

The biggest mistake

Don't try too hard. Most successful reels are surprisingly simple. People do not need cinematic drone footage, aggressive transitions, loud music, or orchestral flamboyance over slow motion cantering. In fact, overly-polished horse content can be quite annoying to the viewer.

You already have the content. But if you do not know what to film, how to structure it, or what wording to use, AI solves this fast.

Give me 10 short Facebook reel ideas for a horse breeding programme focused on nutrition, quiet handling and scientifically sound young horse development.

Suddenly you have structure. Short, clear, one message only, calm pacing, real footage, realistic presentation. That is enough. You may think nobody wants to see you feeding horses. Horse people absolutely want to see it, even if only to give themselves reassurance. We are a deeply strange species.

Easy reel ideas you can film today

Start with daily stable routines: feeding time, rug changes, foal handling, grooming routines, or turnout. Move into educational content such as grass management, young horse handling, or why routine matters. Breeding content works beautifully too; mare updates, foal milestones, stallion handling, or bloodline discussions. A short, well-made promotional video of an open day, a stallion display, or a foaling season can do more work than a month of text posts.

Buyer education is consistently popular: what makes a suitable first horse, questions buyers should ask, and why vet checks should be routine.

A good hook, an opening line that makes people stop scrolling, makes a real difference. Short ones work best for attention.

Film five to ten short steady clips, add a little text, and quiet music if you like, keep it under thirty seconds, and post consistently. People massively over-complicate this. Your audience wants credible information, not entertainment, not a Hollywood production.

Give me 20 reel ideas for a small local leather-work and rug repair service that will pick up and deliver.

Write a 30 second Facebook reel script explaining why temperament for amateur riders should be considered more than flashy movement

Write an educational voice-over for a short video showing a young horse learning to load quietly.

Help me create captions for a reel showing daily handling of foals. Include suitable hooks.

Create short text overlays for a reel showing how to purchase and fit a horse rug correctly.

Do not wait for perfection

Horse people often delay posting because the stable is messy, the horse is fluffy, the lighting is wrong, or the video is not quite right. Meanwhile somebody else is confidently posting absolute nonsense filmed vertically through a dirty float window and getting views. Useful beats perfect, always, especially online.

What short videos build

Good horse videos build familiarity, trust, recognition, reputation, education, and connection and support clients better. Over time people begin recognising your horses, your standards, your handling, your communication style, and your values. That matters, because people increasingly buy from services, they feel they already know.

Using AI to plan, and Clipchamp to build

You do not need a video editor, a production team, or expensive software. The most practical approach for horse people is a two-step process. First, use ChatGPT or Claude to write your script, plan your shots, and structure the video. Then use Clipchamp to put it together using your own photos and footage.

Clipchamp is free, has no credit system, and produces videos without a watermark. If you use Windows 11 it is already installed. If you use a Mac, it is available as a free web app through a Microsoft account. It includes templates for vertical reels, square posts, and standard horizontal video, along with text to speech voice-over, automatic captions, and a library of background music.

The following examples are built around an open day event, a short promotional video for your website or YouTube, and a Facebook reel for social media.

Video 1: Open day promotional video

Suitable for website, YouTube, Facebook page, and email newsletter.

Step 1: Plan the script with AI

Use the prompt below in ChatGPT or Claude. Fill in the bracketed fields with your own details before pasting.

Step 2: Gather your footage

Using the shot list AI has given you, take photos and short video clips around the property. Natural light, calm horses, and ordinary moments work better than staged ones. Aim for more footage than you think you need, since editing down is always easier than realising you do not have the shot you wanted.

Step 3: Build the video in Clipchamp

Open Clipchamp and choose a horizontal template suited to YouTube or Facebook. Import your photos and clips in the order suggested by your shot list, and drag them onto the timeline.

Add the narration from your script either by recording your own voice or by using Clipchamp's text to speech feature, which will read your script aloud in a natural voice.

Add text overlays for the event name, date, and location using the text tool. Choose a relaxed, rural music track from Clipchamp's library and adjust the volume so it sits quietly under the narration.

Step 4: Review and adjust

Watch the full video through. Most images only require 3-4 seconds view time. Any longer than that, viewers tend to switch to something else. Check that the narration matches what is on screen, that text overlays are easy to read, and that the pacing feels calm rather than rushed. Trim or extend clips on the timeline as needed.

Step 5: Export and share

Export the finished video. Share it directly to your Facebook page, upload it to YouTube, embed it on your website, or include the link in your newsletter. A reel between 12 – 20 seconds, and a video from sixty to ninety seconds is a practical length.

Video 2: Facebook reel for the Open Day

A Facebook reel is short, vertical, and designed to be watched on a phone. The goal is not to explain everything; it is to create enough curiosity that people click, share, or save the event date. Thirty seconds done well beats ninety seconds of information overload.

Step 1: Plan the reel with AI

Advanced Prompt:

Theme: Facebook reels, Instagram reels, and TikTok.

Suitable for: Any person wanting to promote their Open Day.

Write a script for a 30 second vertical Facebook reel promoting a [Welsh Cob] stud open day. Tone: warm, energetic, genuine. Audience: horse owners, families, and people curious about Welsh Cobs. The reel should create excitement and urgency without being over the top.

Key messages to include, one or two per scene: meet four Welsh Cob foals born this season, watch stallion displays morning and afternoon, see a breaking-in demonstration showing real horsemanship step by step, farm tours every two hours, questions welcome with students and breeders on hand all day, gates open 10am, date: [INSERT DATE], location: [INSERT SUBURB/TOWN, STATE], book your place: [INSERT PHONE/EMAIL/WEBSITE].

Structure: open with the strongest image or statement in the first three seconds, as this is what stops the scroll. Suggest short text overlays for each scene, five words maximum per overlay. End with the date, location, and contact details on screen. Suggest background music that is upbeat but not frantic, something that feels rural and positive.

Step 2: Choose your strongest vertical photos

The first image is what stops someone scrolling. Use your best foal photo, your most striking stallion shot, or a handling moment that shows the quality of your program. Portrait orientation photos, taller than they are wide, fit the vertical reel format without cropping.

Step 3: Build the reel in Clipchamp

Open Clipchamp and choose a vertical 9:16 template, suited to reels and Stories. Import your photos and clips in the order from your script, placing your strongest image first.

Add the short text overlays from your script to each scene. Keep these brief and large enough to read at a glance, since most people watch reels with the sound off. Add your background music from Clipchamp's library. Play the preview and watch the first three seconds specifically. If the opening does not immediately show something interesting, move a stronger image to the front. You have three seconds to earn the next twenty seven.

Step 4: Export and post

Export the video and confirm it is in vertical format before posting. On Facebook, use the reel option as reels are distributed more widely by the Facebook algorithm. Post it again three to seven days before the event for maximum reach and *pin it* to the top of your page.

Advanced prompt:

Theme: Six-part educational video series

Suited to: trainers, breeders, equine professionals with specialist knowledge

Act as a content strategist and scriptwriter specialising in educational video for the equestrian industry. I want to plan and script a six-part video series on [your chosen topic, e.g. starting young horses / foal handling from birth / fitness for amateur competition horses / the principles of natural horsemanship / rehabilitation after tendon injury].

My audience is [describe e.g. amateur horse owners with limited experience / serious competitors / new horse owners / breeders handling foals]. My platform is [YouTube / Facebook / Instagram]. My video length target is [e.g. 5 to 8 minutes per episode / 60 to 90 seconds per reel].

For the full series, provide:

1. Six episode titles that form a logical progression from foundation to advanced.

2. Three to four key talking points for each episode.

3. A suggested visual or demonstration for each episode.

4. A hook for the opening of each episode, the first line that makes a viewer stay.

5. A suggested posting schedule across six weeks.

6. A series trailer description of 60 seconds that introduces the whole series and tells viewers why it is worth watching from the beginning.

Podcasting and audio content

There is a strong argument that podcasting is more accessible for horse people than video. You do not need good lighting. You do not need to be dressed up. You do not need to be on camera at all. You need a phone, a quiet space, and something worth saying, which most experienced horse people have in abundance.

Podcasting is growing rapidly among horse educators, trainers, breeders, and breed societies, and the barrier to entry is lower than most people assume. A basic episode can be recorded on a phone, lightly edited, and published to a platform that makes it available worldwide within the same day.

Why audio works well

Horse people listen while they work. Feed runs, early morning stable checks, fence walks, and road trips, are all times when someone might comfortably put in an earphone and listen to a podcast about horse health, training, breeding, or management. Audio content reaches people during hours when they cannot read an article or watch a video, which is a genuine advantage.

Advanced prompt:

Theme: Planning and launching a horse podcast

Suited to: horse educators, experienced breeders, trainers, authors, and anyone with specialist knowledge to share

Act as a podcast content strategist with experience in the equestrian industry and audio content creation. I want to start a podcast aimed at [describe your audience, for example: first-time horse owners / experienced breeders / amateur competition riders / horse business owners] and I want help planning it properly before I record a single episode.

Please ask me the following questions one at a time and wait for my answer before continuing:

What is your area of expertise, and what do you know that your audience genuinely seeks to hear?

Who is your ideal listener, and what problem are they trying to solve?

How long would you like each episode to be, and how often do you want to publish?

Do you plan to record solo episodes, interview guests, or a combination of both?

Do you already have an email list or Substack following you could launch to?

Then please provide a podcast launch plan: show title, description, five episode ideas, a full structure for episode one, a reusable show notes template, and a quick three-step publishing process for a complete beginner.

Open day brochure

A printed or downloadable brochure is a natural companion to the video. Use the following prompt in ChatGPT or Claude to generate wording, and format it, or format in Canva.

Advanced prompt:

Theme: Open Day Brochure:

Suited to: Any horse person, club member, horse stud, or business wanting to increase their visibility, attract new patronage, and present their horses, facilities, and program professionally to the public.

Create a one-page open day brochure for a Welsh Cob stud. Include a warm welcome paragraph, the day's programme with times, directions, contact details, and a booking reminder. The tone should be professional, friendly, and genuine. Avoid corporate language.

Details: Stud name: [INSERT]. Date: [INSERT]. Location: [INSERT]. Contact: [INSERT PHONE/EMAIL/WEBSITE]. Programme: 10am gates open, foal visits all day, foal handling demonstrations, stallion display morning and afternoon, student-guided Q&A throughout, breaking-in demonstration, farm tour every two hours (30 mins), as single A4 page with clear headings.

Suggest a follow up report with images and captions to submit to a breed newsletter or local paper.

Video Plan: Based on the above, suggest a video structure for the opening shot description, event highlights in sequence, closing call to action with contact details and date.

- *Write clear narration for each section. End with a strong booking reminder.*
- *Suggest background music style, calm and rural, not dramatic.*
- *Suggest images or video shots to use, best length, text overlays and hooks.*

Ask me questions as you go along. Then give me step-by-step instructions to create it in ClipChamp.

This is what it gave me at the first attempt, not perfect, because it has given me a pinto in the image. Obviously you should use your own photos where possible, or change to the colour of an actual Welsh Cob. You might like slightly larger fonts too, or some of the 'clutter' removed.

Substack as a starting point

Substack, which many readers already know as a newsletter and writing platform, has a straightforward embedded video and audio-creation feature which is quick to use, and can be easily uploaded to YouTube, LinkedIn, Facebook, or Instagram..

Within Substack you can record and publish audio episodes directly alongside your written content. Subscribers receive both in the same place. This means you can write a newsletter about a topic one week and publish a short audio conversation about the same topic the following week, building a consistent presence across both formats, without needing to manage multiple platforms.

With a single click, you can upload your video to YouTube from the platform itself.

The advantage of Substack over standalone podcast platforms is that it is built around your email list. Every subscriber is on your list. When you publish something new, whether written or audio, they receive it directly by email. You are not dependent on a platform algorithm deciding who sees your work, because it goes directly to the people who have chosen to hear from you.

That ownership of your audience is extremely valuable, particularly for horse educators and writers building a long-term readership.

The platform itself costs you nothing to use. If you wish to make podcasting a regular feature, for a small fee, Fiverr can also get you a professional intro/outro clip compatible with your branding.

Before recording, AI can help you plan episodes, structure your talking points, write an introduction, and prepare questions if you are interviewing a guest. A well-structured episode always sounds more confident than an unstructured one, and preparation does not need to take long with AI assistance.

After recording, AI can help you write show notes, create a transcript summary, draft a promotional post for social media, and suggest a title and episode description that will attract listeners. You can also do live recordings.

Help me plan a 20-minute podcast episode about the importance of what to look for in young horses for sale. I am an experienced trainer speaking to an audience of first-time horse buyers and amateur owners. Suggest an episode structure with an opening hook, three or four main talking points, and a practical takeaway at the end.

I have just recorded a podcast episode about [topic]. Here are my rough notes from the conversation: [paste notes]. Please write professional show notes of around 150 words, a one-sentence episode description for the platform listing, and three social media captions I can use to promote the episode.

If podcasting interests you, start with a single episode on a topic you know well and could talk about for twenty minutes without preparation. Record it on your phone using the voice memo app, listen back once to check the audio quality, and publish it through Substack, or by email to your existing list. Ask your subscribers what they thought and whether they would like more.

That is all that is required to begin. Everything else, better microphones, editing software, wider distribution, comes later if the response is encouraging.

Stop press: as this book goes to launch, Substack is rolling out more of the features found in Kit. It is rapidly becoming a "one platform does all."

The Equine Advantage

For those of you who would like to learn more about social media video reels for beginners, especially for Instagram and TikTok, I can do no better than to recommend Jay Hunt's wonderful **The Equine Advantage** courses for people using their smartphone.

They are real value for money as she explains every step in detail. Additionally, she provides a series of 100 quick, copyright-free reels you can adapt for your own use.

You can find her here:
https://hub.equinevideopro.com/

You can check out Substack here:
https://substack.com

***

Asking AI on your phone for ideas, but beware, AI gets horses wrong in ways that only experienced horse people would notice.

Chapter 20

Digital opportunities

Turn experience into greater reach

What feels ordinary to someone who has spent decades around horses often feels incredibly valuable to a beginner.

Reading horse behaviour, introducing young horses to handling, understanding bloodlines, nutrition, pasture management, buyer selection, stable organisation, recognising confidence issues early, are highly specialised skills, and because experienced horse people live around this knowledge every day, they may underestimate how valuable it is.

This matters, because the horse world is entering a period where educational communication may become just as valuable as the horses themselves. And AI is making that far easier to build.

Advanced Prompt:

Theme: Research resources for continuing a young horse's training post-breaking

Suited to: horse owners, first-time young horse producers, amateur trainers, pleasure and competition riders.

Act as an experienced equine training consultant with broad knowledge of young horse development, progressive training methods, and [Australian] equestrian resources. You are going to help me research and compile a personalised resource list for continuing the training of a young horse that has recently completed an initial breaking-in period.

Before building the resource list, ask me the following questions one at a time and wait for my answer before proceeding:

1. *What breed, age, and sex is your horse, and how would you describe*

its temperament and the trainer's notes from the breaking-in period. Were there any concerns or standout qualities noted?

2. *What is your ultimate goal for this horse: pleasure riding, trail work, campdrafting, dressage, showjumping, eventing, showing, or something else, and what is your realistic time-frame for reaching that goal?*
3. *What is your own experience level as a rider and handler, and are you planning to continue the training yourself, use a trainer, or a combination of both?*
4. *What facilities do you have access to: arena, round yard, open paddocks, trails: and do you have regular access to an experienced person on the ground to assist?*
5. *Where are you located: state and general region, so I can search for local clinics, training days, and professionals who advertise services in your area?*
6. *Are there any training philosophies you are drawn to or wish to avoid: for example natural horsemanship, classical dressage, stock horse methods. Are there any trainers whose work you already follow?*
7. *What is your budget for training resources, clinics, and professional help over the next twelve months?*

Once I have answered all questions, compile the following as a structured research list:

1. *YouTube channels and video resources from recognised trainers relevant to my horse's breed, temperament, and intended discipline*
2. *Books, manuals, and online courses from respected trainers that suit my experience level and goals*
3. *Australian clinicians and educators known for young horse development and my chosen discipline, including how to find their clinic schedules*
4. *Local clubs, equestrian centres, and training days in my region that would provide appropriate experience for a young horse*
5. *Local professionals, trainers, coaches, and educators, who advertise services relevant to young horse continuation training in my area, and how to find and evaluate them*
6. *A suggested twelve-month training outline with progressive milestones appropriate to my horse's age, the breaking foundation already in place, and my ultimate goal*

7. *Present the final list in clear sections with notes on why each resource or step is recommended for my specific situation.*

You can see from this prompt the kind of direction horse people naturally seek. But very little of it exists as structured educational material.

Many excellent instructors and horse people know extraordinary amounts, explain things beautifully in conversation, and share advice freely and generously. Meanwhile the internet fills with oversimplified advice, confident misinformation, unrealistic training content, and performative expertise that sounds polished but lacks depth.

This creates enormous opportunity for thoughtful experienced horsemen providing practical guidance. The opportunities here are larger than most people realise. Buyer education guides, foaling preparation checklists, feeding planners, stable management templates, breeding philosophy downloads, young horse handling guides, prompt libraries, horse care journals, video workshops, webinars, and membership communities are all possibilities.

And importantly, these products help horses indirectly by improving owner knowledge.

Help me create a two hour online workshop teaching horse owners how to prepare young horses for successful transitions into new homes. Include lesson structure, worksheets, discussion topics, and practical exercises.

Create an outline for a webinar teaching horse owners how to recognise pain and discomfort in their horses more effectively.

Prompts like these can give you the confidence to finally teach knowledge you have carried quietly for years. Your prompt libraries are becoming valuable tools, but the value still comes from human experience, not the technology itself. Responsibility does not diminish with better tools. It grows.

Create a beginner friendly educational handout explaining normal developmental stages in young horses from weaning to early ridden work.

Help me create a printable management guide for first time broodmare owners including feeding, vaccination, worming, dental routines and foaling preparation.

Membership communities

Horse people can often feel isolated, particularly those in rural areas, though they never complain about it.

They are quietly managing a lot. Small farms running on tight margins, financial pressure that never quite goes away, and the emotional weight that comes with being responsible for animals around the clock.

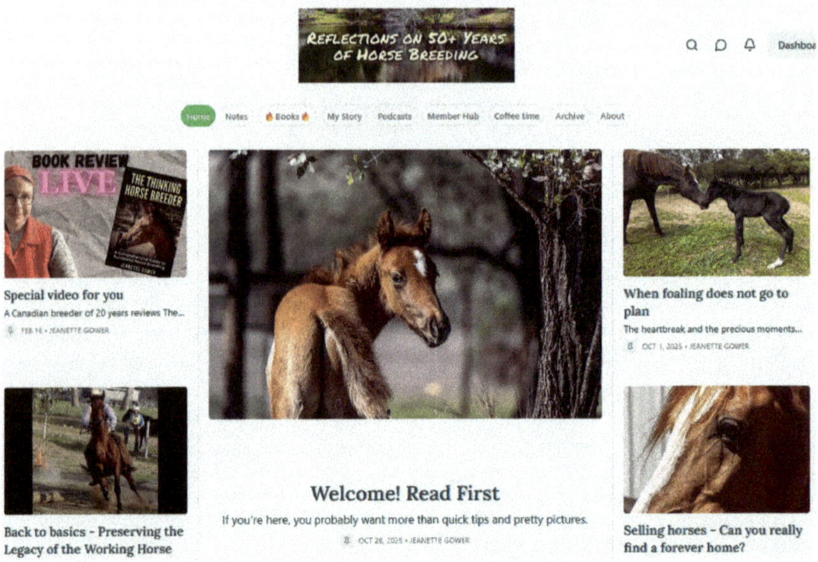

My Substack community website

Subscribe

Tired of theoretical advice & generic tips? Ask questions. Get insights, practical advice, resources, & inspiration. Master the art, science & mindset for excellence, from a seasoned horse breeder who's been in the trenches.

https://jeanettegower.substack.com/

Add injury stress, information overload, and the sheer constancy of the work, and it's no wonder so many feel stretched thin.

That's exactly where a regional AI-supported membership community can make a real difference. Think monthly workshops, ready-to-use prompt libraries, editable templates, breeding discussions, and solid educational resources, all organised so knowledge is easy to find when you need it, not buried in a scroll of old posts.

A community doesn't have to be complicated. A Facebook Group, (public, private, or secret depending on the topic) can be remarkably effective. Take something as specific as senior women who want to keep riding but find incontinence a genuine, practical barrier. That's not a conversation most people want to have in a public forum, but in a closed, trusted group it becomes exactly the kind of honest, helpful exchange that changes lives.

Take my community group on Substack. It is an educational platform, but it also helps me promote my books.

One caution: Some people become so excited by online opportunities that they lose connection to the actual horses themselves. Educational products should always feel practical, grounded, realistic, and experience-based, not internet theory wrapped in motivational language.

Preserve knowledge for future generations

Experience that is never written down tends to disappear with the person who held it. That is the real reason I have written my books. I wanted my family to have access to knowledge I gathered over sixty years; knowledge that took a lifetime to learn and I hope will outlast me.

AI is opening entirely new opportunities for horse people to teach, organise, share, document, and build communities around practical knowledge. Perhaps one of the most valuable things experienced horse people can do now is preserve what they know before it disappears quietly, so that future generations learn from lived experience rather than algorithms.

I have a lot of ideas, so what next?

You do not have a content problem. You have a structure problem.

If you have been creating posts, recording videos, writing articles, starting newsletters, and building an audience, yet somehow that effort is still not producing the engagement you hoped for, the answer is rarely to create more.

Creating more can make things worse. You end up with dozens of half-finished ideas, offers nobody sees, content nobody remembers, and a growing sense that you are working harder than ever for very little result.

What if the answer isn't creating more?

What if the answer is building more from less?

Advanced Prompt:

Theme: Building a paid knowledge product from existing expertise

Suited to: experienced practitioners, therapists, equine body workers.

Act as a digital product strategist with experience in education, professional services, and the equestrian industry. I have been working as a [profession] for [X years] and I have expertise in [describe your specialist area e.g. remedial farriery / starting horses with trauma histories / breeding for specific traits / equine sports massage for competition horses].

I want to turn my existing knowledge into a paid digital product, something that generates income without requiring my direct time for every sale. I am not a technical person and I do not want to build a complicated platform.

Please do the following:

1. Based on what I have described, suggest three possible digital product formats suited to my expertise and audience, for example a short practical guide, a video masterclass, or a reference document, and explain the realistic effort required for each.

2. For the format you think is strongest, outline a complete product structure: a title, an introduction, the core content sections with a brief description of each, and a conclusion.

3. Identify the single most valuable piece of knowledge I could share that my target audience would pay for, the thing that saves them time, money, or mistakes.

4. Suggest a realistic price point and where I could sell it without building my own website.

5. Write the first paragraph of the introduction in my voice, based on what I have told you about my background.

Advanced Prompt:

Theme: Personal AI integration audit

Suited to: any horse professional wanting to use AI strategically rather than randomly

Act as a business efficiency consultant with deep experience in AI tool integration for small businesses and professional sole traders. I am going to describe my current working week in detail, what I do, how long each task takes, what I find draining, and where I lose time. I want you to identify the highest-value opportunities for AI to help me, ranked by impact and by how easy they would be to implement.

Here is my typical working week: [describe honestly. Include administrative tasks, client communication, marketing, record-keeping, physical horse work, travel, anything that takes your time. Include tasks you dislike, tasks that repeat, and tasks that feel disproportionate to their value].

Once you have read it, provide:

1. The five highest-value opportunities for AI assistance in my current workflow, ranked from highest impact to lowest. For each one, tell me: what the task is, how AI would help, which specific tool you would recommend, and a realistic estimate of how much time it could save per week.

2. The two or three areas where AI would not help and I should not waste time trying.

3. A suggested starting point. The single first task I should tackle this week to begin integrating AI in a way that will stick.

4. A six-week plan for building AI into my workflow gradually, one area at a time, without disrupting the work that is already running well.

How to promote your product?

Once you have built your product, the next question is how to get it into people's hands without needing a complicated website or a large technical budget.

The good news is that plenty of platforms already exist to do exactly that. A downloadable guide or checklist can be sold through Gumroad, which is simple to set up and requires no technical knowledge.

Printed booklets and physical products can be listed on Amazon through Kindle Direct Publishing, which handles on-demand printing and postage on your behalf.

If you want someone else to format or design your product professionally before you sell it, Fiverr connects you with affordable freelancers who specialise in eBook layout, cover design, and document formatting.

Online courses and video workshops sit comfortably on Teachable, Thinkific, or Kajabi, all of which handle payments, delivery, and student access automatically. If you prefer a community model, Substack supports paid newsletters and subscriber communities with minimal setup.

Facebook Groups remain one of the simplest ways to build a free or low cost membership community around a specific topic. And if you are starting out and want to test whether people will pay for your knowledge before building anything at all, a PayPal link and a PDF sent by email is a perfectly legitimate first step. The platform you use is far less important than the knowledge inside the product. Start simple, deliver genuine value, and scale from there.

Advanced Prompt:

Theme: The Knowledge Ladder

Suited to: hobbyists with knowledge to share, and horse industry professionals such as equine nutritionists, dental therapists, farriers, acupuncture practitioners, trainers, and breeders.

Act as a digital education strategist with experience working with sole traders, small business owners, and specialist professionals in the equestrian industry. I have knowledge and experience in [describe your area, for example: equine nutrition for performance horses, natural hoof care, starting problem horses, breeding for temperament, or general horse ownership and management]. I want to understand how I could turn that knowledge into a range of products and services at different price points, starting with something basic and low cost, and building toward higher value offerings over time.

Please build me a Knowledge Ladder with five rungs, moving from free or low cost at the bottom to premium at the top. For each rung, tell me: what the product or service could be, a realistic price range I could charge for it, where I could offer or sell it without needing my own website, how much time and effort it would realistically take to create, and who it is best suited to.

For each rung, also show me two or three branches, meaning other directions that same product or idea could grow into with a small amount of extra effort. For example, a free checklist might branch into a paid expanded guide, a short video series, or a live question and answer session. A paid ebook might branch into an online course, a printed booklet sold at events, or a monthly membership with ongoing content. Show me how one piece of knowledge can travel in several directions rather than stopping at a single product.

Then help me evaluate which rung to start with based on my current situation by asking me three questions: How much time do I have available each week to work on this? Do I prefer writing, speaking, or teaching in person? And am I starting from scratch with an audience or do I already have followers or clients who know my work?

Once I answer those questions, recommend my best starting point, explain why that rung makes sense for where I am now, outline a simple pathway for moving up the ladder over the next twelve months, and suggest which branches are most worth exploring first, based on my answers.

Digital products and e-guides on GumRoad: https://jeanettegower.gumroad.com /? Readers of this book can get a 20% discount on all products by entering code CHALANI20

Self-Publishing

Would you like to do something serious? Have you thought about becoming an author? I am not here to suggest you do an AI e-book, nor something created by AI. AI generated books are easy to pick and platforms are increasingly flagging and removing books which are obviously written by AI.

That doesn't mean you can't use AI to assist you with say, a book outline, chapter headings and editing, after you have written a draft manuscript. It takes genuine hours of work to launch a book that really interests readers, with some hope of a monetary outcome. This is not for the faint-hearted, let me tell you.

It is a huge learning curve. I'd suggest that anyone seriously interested in authoring a book for the wider market, first subscribe to some Facebook communities, such as *Self-publishing Support Group*, *Self-publishing Australia*, and *Breakthrough Publishing*. Learn about covers, printing on demand, formatting and the like. This book, and all my books have been formatted on Atticus, a great tool for authors.

Then you also need a marketing plan, leading up to and following the launch. This is as much work as writing the book itself.

Book marketing:

I recommend Rachel Anderson for that role. She can set up launch optimisation, ad campaigns, author profile optimisation, reader love campaigns, and book recommendation campaigns, just to name a few.

Rachel Anderson, Book Marketing Specialist https://racheal-anderson-rabs.netlify.app/

Search terms (SEO), keywords, taglines, hashtags and llmxt:

Search engine optimisation, or SEO, is about helping people find your website or content when they search online, for example when someone types "Welsh A breeder Victoria" or "quiet trail horse for sale near me" into Google. Using relevant words naturally throughout your website, blog posts, and image descriptions, the kind of words a potential buyer or client would search for, makes it more likely your business appears in those results rather than buried on page five.

Keywords are the specific words or short phrases people type into a search bar. A tagline is the short, memorable line that sums up what you offer, such as "Quiet, honest trail horses for the everyday rider," and ideally contains a keyword or two without sounding like a search term. Hashtags work a little differently again; they are used within social media posts rather than websites, and help your content appear when someone searches or browses a topic on platforms like Instagram or Facebook. A post about a young Welsh Cob for sale might use hashtags like #WelshCob or #TrailHorseForSale so that someone scrolling that topic finds it.

Search itself is changing. Increasingly, people are not typing a search term into Google at all. They are asking ChatGPT, Claude, or Google's AI Overview a direct question, such as "who breeds quiet trail horses in Victoria," and receiving a written answer rather than a list of links. This is sometimes referred to as llm optimisation, or llmxt, and it works differently to traditional SEO.

Rather than ranking for keywords, your website needs to contain clear, well-structured information that an AI can read, understand, and confidently summarise when answering someone's question. Clear "About" pages, realistic horse descriptions, and straightforward writing, the same qualities this book has encouraged throughout, are exactly what helps you here as well.

AI can help with this by suggesting natural phrases to include in your writing, by generating image descriptions (called alt text) that describe what is in a photo for both search engines and visually impaired visitors, and by reviewing your existing website pages to suggest where wording could be adjusted to include useful search terms, without sounding forced or repetitive. (See also page 157)

Advanced prompt:

Theme: Internet search visibility

Suited to: any horse business owner wanting to improve how easily they're found online, whether through traditional search engines or AI tools like ChatGPT and Google's AI Overview

Act as an SEO specialist with experience in small business and local search. I run a [describe your business, e.g. Connemara stud / riding school / equine therapy practice] based in [location]. My website covers [list main pages or topics, e.g. horses for sale, about us, services, contact].

Suggest 15 to 20 keywords and phrases that potential clients or buyers in my area might search for when looking for a business like mine, including broad terms, location-specific terms, and more decision-ready phrases. For each one, suggest where it could naturally be included, such as a page title, heading, image description, or body text, without sounding forced or repetitive.

Also suggest a short tagline that sums up what I offer using one or two of these keywords naturally, five hashtags suited to social media posts, and two or three sentences for my website that would help an AI tool such as ChatGPT or Google's AI Overview accurately describe my business if someone asked it directly.

Day's end at Chalani: leaving the chaos behind. When AI takes care of the noise, you get to focus on what you came here for.

Chapter 21

Ethics of AI in the horse world

Looking ahead

Every generation of horse people eventually faces the same question. How do we move forward without losing the values that matter most?

The horse world has always evolved. Feeding practices changed. Veterinary care changed. Breeding methods changed. Training philosophies changed. Communication changed. Now AI is arriving quietly, and is suddenly everywhere.

So how do we use modern tools responsibly without losing authenticity, integrity, and good horsemanship? This is not really a technology conversation. It is an ethics conversation.

Artificial Intelligence is neither automatically good nor automatically dangerous. The same was said when personal computers arrived. A float can transport horses safely or dangerously. A breeding programme can improve horses thoughtfully or recklessly. AI works the same way: the impact depends entirely on the person using it, and whether they keep applying judgement and integrity.

Appearance and reality

The growing gap between appearance and reality is one of the biggest risks ahead. AI makes it easy to create polished nonsense.

A person can generate professional articles, sophisticated advertisements, polished branding, and confident-sounding explanations without decades of lived experience underneath them. Good presentation once reflected years of knowledge and reputation. That connection is becoming less reliable, so the horse world will need to become more thoughtful about what credibility actually looks like. Good branding should support genuine knowledge, not replace it.

The same concern applies to photography and video. AI generated footage has reached a level where fake horses, fake properties, and fake handlers can be created from still photographs and made completely convincing. This is already being used to scam buyers. If you are buying a horse and have not seen it in person, or via a live unedited video, you have not seen the horse at all. (See example: https://www.facebook.com/reel/103472569555196)

Once fake horses, altered conformation, or manipulated presentation become normalised, viewers will rightly question everything they see. Everybody loses, especially the horses.

Polished language can hide a lot

AI is remarkably good at rewriting wording and softening language. Used properly, this can help breeders explain suitability clearly and reduce misunderstandings.

But there is temptation too. A difficult horse can suddenly sound "spirited." Poor behaviour becomes "quirky." Unsuitable placements become buyer misunderstandings. Ethical advertising has always required honesty, and AI does not remove that responsibility. If anything, it increases it. The biggest hurdle is not who uses AI, but who uses it responsibly.

More content is not always more value

Once content becomes easier to generate, there is a temptation to create more of everything: more posts, more slogans, more experts. But volume is not the same as value. Thoughtful communication becomes increasingly valuable precisely because so much online content is growing louder and more artificial. Truly knowledgeable horsemen are remarkably quiet communicators. Experience tends to create steadiness rather than noise.

Advanced prompt: find where your own marketing misleads

Suited to: any horse person serious about ethical communication

Act as an ethical communication auditor with expertise in the horse industry. I am going to share examples of my current marketing and communication [insert website copy, social media posts, or horse advertisements]. Your job is not to be kind. Identify, specifically and honestly, where my communication might be misleading, incomplete, exaggerated, or where it creates expectations reality may not meet.

[Paste your examples here] For each piece, tell me:

- *What claims or implications are made that a buyer or client might rely on.*
- *Which claims are fully supported by the facts as presented.*
- *Which are vague, or rely on the reader not asking follow-up questions.*

- *Where a reasonable person might feel misled after engaging with my business based on this communication.*
- *A suggested rewrite of any flagged passage, keeping the positive qualities while removing or qualifying the misleading elements.*

Looking ahead: developments underway

Some of what follows is not yet widely used in the horse industry, but is developing quickly enough that readers should know it exists before it arrives at their own gate.

AI video generation from text or photos is advancing rapidly. Tools such as Sora, Runway, and Veo can generate short clips from a written description or a single photograph. The line between real and generated footage will blur further, making authenticity more important, not less.

Talking avatars create a digital version of a person that can speak a script in their voice, with lip syncing and natural movement. The technology has improved enormously, but still tends to look slightly off in ways viewers notice even if they cannot say why. An avatar reading a script sits uneasily alongside everything this book has said about sounding like yourself.

Real-time translation is becoming good enough for genuine conversation, not just rough text translation. For breeders selling internationally, this could remove a significant barrier without a third party involved.

AI-assisted veterinary triage tools are emerging, designed to help owners describe symptoms and gauge urgency. Anything touching welfare decisions belongs in the same category as everything else this book has said about AI and horse health: useful for understanding, never a substitute for veterinary care.

AI note-taking tools that listen to a lesson or consultation and produce a summary afterward could give instructors a simple way to follow up with clients without extra admin.

Search engines and GEO

"SEO is dead. GEO is the future. AI killed search; better learn the new rules or get left behind."

Most of us just want to ride or be with our horses. We don't want to decode five new acronyms every few months (SEO, GEO, AEO, LLMSEO). We want the thing we posted to get found by a human who needs it. Enough!

GEO stands for generative engine optimization. Getting your work cited inside AI answers, such as Google's AI Overviews or a ChatGPT reply. GEO decides whether these tools quote your work back to someone searching. The reassuring news, straight from Google, is that good SEO is good GEO; the same qualities that always made writing trustworthy are what AI now looks for.

Google tells creators to move away from generic content and share non-commodity content, work only you could write. A basic "10 tips" post is easy for AI to generate and easy to ignore. A first hand account, drawn from 30 years in the saddle, of why a specific approach worked or failed for a real horse, is what gets noticed and cited.

Structure your website well: give each page a clear, front-loaded title and description. Use real headings, ideally posed as a question, with the answer directly underneath. Keep the site fast and mobile friendly. Group content around two to three core topics and link related pages together, building topical authority.

For Facebook and social media posts, your opening line is both the hook and summary, so lead with the point. Post consistently around your core themes, so readers and AI systems alike build a clear picture of your expertise.

None of this needs shortcuts or special files. Focus on clarity, genuine experience, and a real voice. That is what gets found and quoted.

> Review this website page about [topic, e.g. laminitis prevention in horses / paste actual page text]. Rewrite it to answer the reader's main question in the first sentence. Add clear H2 headings phrased as questions, with the answer directly below each one.
>
> Suggest a specific, front-loaded title tag and meta description under 160 characters. Point out anywhere the content is generic and suggest how I could replace it with a first-hand example from my own experience. Keep language plain and avoid keyword repetition.

Quick checklist (no AI required)

- Does the first paragraph answer the reader's question directly?
- Does every heading pose a question, with the answer right underneath?
- Is there at least one example from real experience, not just general advice?
- Is the page fast to load, with images compressed?
- Does it link to two to three related pages on your site?

Other technology

Wearable and sensor technology for horses, heart rate monitors, movement trackers, and girth-mounted sensors, already exist. AI is increasingly used to interpret the data they collect: early indicators of lameness, fitness trends, and recovery patterns may become visible long before they would be obvious to the eye.

Phone AI is also moving toward recognising individual animals rather than just identifying that a photo contains a horse. For breeders managing large numbers of horses over many years, searching photos by which specific horse appears in them could save enormous amounts of time.

None of these tools are mentioned as recommendations. They are mentioned so that, as they arrive, and some already have, they do not feel unfamiliar or overwhelming. The questions this book has asked throughout, does this help the horse, does this support good judgment, does this build trust, will still be the right questions to ask of whatever comes next.

Advanced prompt: evaluate a new AI tool before adopting it

Suited to: anyone considering a new AI tool for their horse business

Act as a thoughtful advisor helping a horse business owner decide whether a new AI tool is worth adopting. I am considering using [describe the tool and what it claims to do]. My business is [describe your business briefly].

Please ask me questions one at a time to help me think this through: what problem this would solve for me, what it would cost in time and money, what could go wrong if the tool gives me incorrect information, whether my clients or buyers would want to know I am using it, and whether there is a simpler way to achieve the same result without it.

Once I have answered, give me a clear summary of whether this seems worth trying, and if so, suggest one small way to test it before committing fully.

The future belongs to adaptable horse people

Horse people adapt constantly, without really thinking about it. Seasons change, horses change, markets shift, weather wrecks the best laid plans, and somehow you adjust and carry on. That same adaptability is what AI now asks of us.

No algorithm develops 'feel'. No chatbot reassures a frightened horse through body language and timing. Horses still demand authenticity, and they still expose inconsistency, impatience, and ego, however polished your online presence.

The danger was never AI itself. It is poor judgement, amplified. Used well, it helps you communicate and organise. Used carelessly, it creates noise very quickly. The difference is always *you*.

I hope this book has given you plenty of ideas, and the confidence to try them.

– Jeanette
Always with the horses in mind.

A simple question about a foal's conformation sparked a much bigger discussion about expertise, trust, and the role of AI in the horse industry.

Can you compare a horse to a breed standard?

Recently , a horse breeder dipped a toe into the AI world. Not with a business plan or a grand announcement. She just asked a question:

"Can you compare a horse with a breed standard?"

The horse was a young foal. The AI assessment was posted publicly and then shared on Facebook.

What struck me immediately was that many people reacted to the idea of AI assessing conformation without looking closely at what had happened.

The original question was broad, and the AI did not immediately launch into an assessment. Instead, it asked for more information. It requested side, front and rear photographs, a head and neck profile, and details about the horse's breed before attempting any comparison.

That detail matters. Whether you love AI or hate it, the system itself recognised that conformation assessment requires more than a single photograph.

It recognised the lack of context and asked for further information, photos, sire and dam.

What people were really arguing about

What interested me most was not the AI's response. It was the discussion that followed.

At first glance the comments appeared to split into two camps. Those who disliked AI and those who did not. But the longer I read, the more I realised that was not really what was happening.

The discussion was less about artificial intelligence and more about *trust*. Who do we trust when evaluating a horse? The experienced breeder with decades of practical knowledge? The judge standing in the middle of the ring? The photographer behind the camera? The breed standard? The horse's performance record? Or an AI model?

Some people clearly objected to AI itself. Their concerns centred around environmental impact, critical thinking, creativity, corporate influence, and the loss of traditional skills. Others were not necessarily anti-AI at all. They rightfully questioned whether conformation is a task that AI can perform reliably.

The strongest observations came from horse professionals

One of the strongest observations came from a professional horse photographer. She explained how easily a horse's apparent conformation can be altered by changing the angle of a photograph, adjusting the horse's stance, moving the head, or using different equipment. A quality horse can look ordinary. An ordinary horse can look exceptional.

That cuts straight to the heart of the issue, because suddenly the AI is not necessarily assessing the horse. It is assessing the photograph.

Several breeders raised a similar concern from a different perspective. More than one person described traveling to inspect stallions they had previously only seen in photographs and videos, only to discover the horse looked very different in person. Different type. Different presence. Different suitability for their mares.

This is not really an AI problem. It is a real-world problem. Photographs can tell us a great deal about a horse, but they can also mislead us, just as any value-judgement can.

What surprised me most, however, was how few people appeared interested in testing the claim itself. Some commenters stated confidently that AI would not work. Others claimed it was already proving useful. Yet only a handful suggested meaningful testing and comparison against experienced breeders knowledge, or performance. The AI tools methodology or scoring system was not questioned.

That is how we discover whether a tool has value. Not through opinion. Through testing.

What this discussion revealed about trust and expertise

Another interesting pattern emerged when people began discussing how they used AI. The most balanced comments often came from those who were already using it, not as a replacement for expertise but as a supplement to it. People described using it to help interpret medical reports, explore saddle fitting options, research biomechanics, and better understand training concepts.

None of these people appeared to be surrendering their judgement to a machine. They were using it as a starting point rather than a final answer.

Other experienced horse people expressed concern that newer breeders may stop learning conformation from mentors, judges, and seasoned breeders. That concern is not irrational. Historically, conformation assessment has been built through observation, mentorship, studying bloodlines, attending shows, handling horses, and learning from mistakes.

At the same time, some of the people using AI were making almost the opposite argument. They saw it as a tool that helped them learn faster, ask better questions, and access information they might not otherwise have found. Interestingly, both groups were defending knowledge. They were simply approaching it from different directions.

If I had to summarise the overall discussion, I would say the majority position was neither that AI should never be used nor that AI should be trusted. Most people seemed to accept that AI may have value as a tool while also recognising the limitations of assessing conformation from photographs. More importantly, most agreed that no report, score, photograph, or computer generated assessment should ever replace experienced human judgement.

The Facebook post started as a simple experiment. What it revealed was something much bigger. Not what people think about AI, but what people think about expertise.

And that may be the more important conversation.

PART IV

Bonus section

Use this section as a resource to consolidate your knowledge.
For readers of this book only, I have compiled a Google Doc, of all the prompts, so you can copy and paste relevant prompts into your own library.
You can find it here:

https://docs.google.com/document/d/1A38Fvx3Ru7u5HLZ33pbGbfqSu9gGiec3/edit?usp=drive_link&ouid=10748786904814232491 9&rtpof=true&sd=true

Bonus Chapter

Practical Prompt Ideas

These prompts are designed to be practical, realistic, and easy to adapt for your own horses, breeding program, riding school, or horse-related pursuits. Simply edit accordingly.

Administration

Create a monthly newsletter template for a horse breeding program.

Please convert this image into text.

Create a checklist for preparing horses and paperwork for transport.

Help me research average sale prices for [breed, age, sex, height, intended use, year] at auction and, where possible, through private sales. Please indicate how you arrived at this information.

Identify health certificate and movement clearance testing requirements for interstate and international transport from [location 1] to [location 2], and how long before movement such tests must be completed. Please refer me to the relevant authority websites for reference.

Business identity

Help me brainstorm branding ideas for a horse breeding program focused on traditional bloodlines and family horses.

Suggest colour palettes for a horse property wanting to feel calm, timeless, and professional.

Create tagline ideas for a horse breeding program focused on temperament and versatility.

Help me write a short brand or mission statement for my horse operation.

Suggest logo concepts that avoid generic horse head imagery.

Buyer communication

Create a buyer information sheet for somebody purchasing their first young horse.

Help me explain my feeding routines clearly for a horse buyer information pack.

Write a polite and professional response declining an unsuitable buyer enquiry.

Educational content

Summarise this veterinary article into practical everyday language.

Create a simple educational handout about introducing young horses to new environments.

Write a beginner-friendly explanation of horse herd behaviour.

Create a short educational article about why consistency matters in horse training.

Please create an alphabetical checklist of native and introduced plants toxic to horses in my area [insert region]. For each plant include a photo, brief description, degree of toxicity, and symptoms. I would like to laminate this for my feed room. [insert any additional requirements]

Please suggest checklists and how-to topics that would be valuable for a Pony Club or children's riding establishment, suitable for round-table discussion on rainy days, with possible discussion questions for each and illustrations where appropriate.

Performance analysis

I am a racehorse trainer evaluating the form of [horse name] ahead of an upcoming race. I have copied the following form data from [insert source]. Please analyse this data and give me a structured assessment based on the following priorities in order: recent race results over the last five starts including finishing position, margin, field size, and class of race; track condition performance across good, soft, and heavy going; optimal distance range compared to today's trip; weight carried and any changes; barrier draw history; sectional times and finishing pattern; days between runs and preparation; gear changes; and current stable form.

Please identify the horse's strongest qualities, any concerns, and give me a plain summary of whether this data suggests the horse is suited to its next assignment. I will supply the form data below. [paste form data]

Research

What do the suffixes after a horse's pedigree name mean in the Australian Stock Horse?

What does "bold type" mean in a Thoroughbred racehorse's pedigree?

Help me research the different types of non-steel horseshoes, their uses, and their suitability for different types of horses.

Identify core and risk-based vaccination requirements for a horse located in [country/state/area]. Briefly explain the disease and the recommended protocol for each.

Search for horse properties or rural acreage matching the following criteria: [insert requirements, for example infrastructure, number of paddocks, house requirements]. Location within [insert radius, for example 50km] of [insert location]. Price range between [minimum] and [maximum]. Property size between [minimum acres/hectares] and [maximum]. Additional requirements: [for example school bus route, bitumen road access, town water]. Please scan real estate and rural listing websites in [insert country/region], including [relevant websites for your area], and return a list of current listings matching these criteria. Present results in price order, with the property address, a brief description, land size, and a direct link to each listing.

Please explain the registration systems for Iberian horses, covering both the PRE (Pura Raza Española) and the PSL (Puro Sangue Lusitano). For the PRE, explain how the Spanish studbook works, what ANCCE is and what it does, and the difference between a PRE and an Andalusian. For the PSL, explain how the Portuguese studbook works, what APSL is and what it does, and the difference between a PSL and a Lusitano.

Please explain the common abbreviations, codes, suffixes, and brand markings that appear in PRE and PSL pedigrees, including the LGN and LGA numbers used in PSL registration, breeding prefixes from the major Portuguese stud farms such as Andrade, Veiga, Alter Real, and Coudelaria, and any grading or inspection designations used in either system.

Routine organisation

Create a seasonal pasture management checklist for broodmares, foals, and spelling horses.

Help me organise feeding schedules for multiple horses with different requirements.

Create a foal development tracking template including handling milestones and behavioural observations.

Help me organise veterinary notes into a simple timeline format.

Please do a nutrition and cost comparison of these feed bag tags [attached] from my local store, and give a summary of value for money based on manufacturer recommendations, my horse's weight, and current feeding regime and workload.

Sales advertisements

Rewrite this horse sales advertisement to sound calmer, clearer, and more professional while still feeling warm and approachable.

Create a Facebook sales advert for a young horse suitable for an experienced home wanting a future all rounder.

Help me create an A4 style sales advert for my [advanced level dressage horse, competitive reining horse] using this photo of him. Do not change the image. Summarize [performance record, training profile, breeding, registration, location and contact etc.] Suggest fonts, colours, text overlay, heading, and captions for the advert.

Social media content

Write an educational post explaining our process for collecting and sending semen.

Help me write a calm and informative social media caption for a young horse in training video.

Help me write three variations of the same post about [topic], one formal, one conversational, and one with a touch of humour, so I can choose the tone that suits my audience.

Help me write a social media post that tells the story behind one of my horses, including their breeding, personality, and journey, in a way that feels genuine rather than promotional.

Suggest five ways I could repurpose one piece of existing content, such as a blog post, article, or educational handout, across different social media platforms without it feeling repetitive.

Help me respond professionally and calmly to a negative comment or review left publicly on my horse page. The comment says: [paste comment]

Compare Instagram, TikTok, Pinterest, YouTube, and Reddit as platforms for horse owners. For each, explain the type of content that performs best, the typical audience, how much time it realistically requires, and whether it suits a small breeding or training operation, a riding school, or an individual horse owner. Help me decide which platform or platforms are worth my time alongside Facebook, based on [describe your situation and available time].

Spreadsheets and systems

Please recommend and compare horse management apps and software suitable for a small property managing broodmares, foals, breaking-in, routine tasks, and veterinary records. For each suggestion, outline the key features, ease of use for non-technical horse owners, and approximate cost. Include what features I should look for before making a decision.

Should our organization consider adding a [chatbot / booking system / online applications / other] to our website, and what could it realistically do for visitors and enquiries? Please explain the types available, what might suit, what the approximate cost and time commitment would be to maintain one.

Please set up a proper tracking spreadsheet for [pasture record and property management / horse buyer contacts and follow-up / competition and show entry tracking / expert interviews and sources / social media content planner for horse topics performing well / equine therapy and rehabilitation progress tracking / horse transport, hay and agistment contacts / maintenance schedules and contractors / emergency contacts.] Please include all relevant columns and explain how to use it effectively.

Help me design a comparison chart for the five most popular vehicles used to tow a two-horse float in Australia.

1. Create five columns, one for each vehicle, plus a final column showing approximate new price range in Australian dollars.

2. Down the left-hand side include criteria covering towing capacity and rating, engine type and transmission, payload capacity, fuel type and economy, stability and safety features, weight of horses being carried and how this affects vehicle choice, type of towing required, suitability by user type, reliability rating, and known problems or recurring issues reported within the last ten years.

3. Inside each cell, give me the key questions I should be asking or specific figures I should be researching for that vehicle against each criterion. At the end, in a short paragraph, list the most useful Australian websites offering vehicle comparison tools, towing guides, and owner review forums with a brief note on what each offers.

4. Using the tow vehicle comparison chart framework we have just created, please now research and fill in the relevant data, figures, and answers for each of the five vehicles across every criterion listed. Where exact figures are not available, give me the best available estimate and indicate how you arrived at it. Flag anything that needs to be verified directly with a dealer or manufacturer before purchase.

Stud

Create wording for a stallion flyer including breed data, service details, breeding highlights, and contact information.

Write an advert announcing that chilled and frozen semen is internationally available for [name/details.]

Build me a financial [model/spreadsheet/checklist] to help evaluate the annual cost of producing a registered [breed] foal to saleable age ([insert age]), compared against likely sale price, based on research into average costs for this type of horse.

I am considering purchasing [chilled/frozen] semen from [stallion name, breed, country of origin] through [supplier or agent name]. Please help me verify the credibility of this transaction by researching whether the supplier or agent is a recognised and reputable operator, and provide a list of questions I should ask the supplier and my veterinarian before proceeding. Please flag any aspect of this transaction that warrants further verification.

I am uploading or pasting a semen analysis report provided by the supplier for [stallion name, breed, country of origin]. Please analyse the key parameters including progressive motility, total motility, morphology, concentration, and post-thaw survival rates, explain what each figure means in plain language, and flag anything that falls below internationally accepted thresholds for transported or frozen semen. [attach or paste report]

Tips for better prompting

I would like help writing [insert task, for example a sales advertisement, a social media post, or a report to a veterinarian]. Please write this in a [insert tone, for example warm and conversational, calm and professional, or friendly but informative] style. Here is the information I would like to include: [paste your notes]

I have written down my thoughts on [insert topic] exactly as they came to me, without worrying about order or wording. Please re-organise these notes into a clear, readable piece that flows naturally and sounds like me. Do not change the meaning or add information I have not provided. Here are my notes: [paste notes]

I find typing prompts difficult or time consuming. Please give me a step by step guide to using voice dictation on [mobile/desktop] to speak my prompts aloud instead of typing them. Include any tips for speaking clearly so the dictation is accurate, and how to review and correct the text before sending it to AI.

Using images

I am uploading a photograph of [horse name/breed/age] taken during extended gallop. Please assess shoulder angle and freedom of extension, hip angle and hindquarter engagement, hock flexion and hind leg reach under the body, knee and fetlock extension in the foreleg, hoof-pastern axis alignment, topline including back length, loin coupling and croup angle, proportional balance between forehand and hindquarter, and neck carriage at the moment of capture.

For each, describe what you can see, note whether it appears correct, and flag anything warranting closer assessment by a veterinarian or equine biomechanics specialist. [attach photo]

I am uploading a square-on conformation photograph of [horse name/breed/age]. Please assess straightness of limbs front and behind, width and depth of chest, symmetry of muscling through the shoulder, hindquarter and topline, and proportional balance. Note anything that appears correct, anything of concern, and anything that cannot be determined from a still image alone. [attach photo]

I am uploading a photograph of [horse name]'s hoof. Please assess the hoof-pastern axis, heel height and balance, any visible flare or distortion, and whether the shoe fit appears appropriate. Note anything that warrants discussion with a farrier. [attach photo]

I am uploading a photograph of a wound or skin condition on [horse name]. Please describe what is visible in plain language to help me communicate clearly with my veterinarian. This is not a request for diagnosis. [attach photo]

I am uploading a side-on photograph of [horse name]. Please estimate a body condition score using the Henneke scale, describe what you can see at each scoring region including neck, withers, shoulder, ribs, loin, and tail-head. [attach photo]

I am uploading an existing photograph of [horse name/description]. Please help me improve this image by [describe what you would like, for example brightening the exposure, removing a distracting background, or making the horse stand out more clearly]. This is for use on [insert purpose, for example a sales advertisement, social media, or a stud profile]. [attach photo]

A reminder on image analysis: A photograph captures one moment and one angle only. AI can assess visible proportions, angles, and alignment from a still image. Use these prompts as a starting point for conversation with your vet, farrier, or trainer, not as a substitute for professional assessment.

Website and brochure wording

Help me improve the wording on my horse property website while keeping it warm and professional.

Rewrite this brochure to sound clearer and more polished without becoming overly sales focused.

Write an About Us page for a horse breeding program focused on our family history with horses and responsible breeding.

Create website headings for a small training establishment, focused on showing quality in-hand horses on behalf of owners.

I have a website for my [horse property/stud/riding school]. I have heard that AI search tools now look for a file called llms.txt on websites to understand what a horse owner or operation does. Please explain what llms.txt is, whether horse owners with a website should consider having one, what it should contain, and how I would go about creating and adding one to my site.

Practical Ways That You Can Use AI

This section provides practical examples that may spark ideas. Not every suggestion will suit every reader. That is perfectly normal. The goal is to identify opportunities that fit your own circumstances.

Administration

- Build business manuals, workbooks.
- Build procedures.
- Create checklists and FAQs
- Create forms and templates.
- Develop record-keeping systems.
- Draft policies.
- Improve workflows.
- Organise digital files.
- Organise documents.

Agistment and boarding

- Build client information guides.
- Build communication templates.
- Create feeding information sheets.
- Create onboarding systems.
- Create welcome packs.
- Develop emergency procedures.
- Develop frequently asked questions.
- Draft boarding agreements.
- Organise horse records.
- Produce newsletters.

Breeding

- Build breeding resource libraries.
- Calculate breeding formulas and genetic information.
- Compare breeding options.
- Create buyer education guides.
- Create mare profiles.
- Develop stallion information sheets.
- Draft breeding newsletters.
- Organise bloodline research.
- Organise foaling records.
- Preserve historical breeding information.
- Summarise breeding articles.

Communication

- Create frequently asked questions.
- Create parent information packs.
- Create welcome packs.
- Draft client emails.
- Draft event announcements.
- Draft follow-up messages.
- Improve horse sale advertisements.
- Improve newsletter content.
- Write lesson reminders.
- Write website content.

Education

- Build educational checklists.
- Create resource libraries and book lists.
- Create study guides.
- Create webinar content.
- Create worksheets.
- Develop learning plans and teaching material.
- Develop workshop outlines.
- Generate discussion questions.
- Summarise research.

Events

- Build volunteer information packs.
- Create educational resources.

- Create follow up communications.
- Create registration and entry information.
- Create reminder emails.
- Develop post event surveys.
- Draft event schedules, and programmes.
- Draft sponsorship proposals.
- Organise event workflows.
- Produce participant guides.

Horse sales

- Build sales information packs.
- Build sales workflows.
- Create buyer preparation guides.
- Create comparison tables.
- Create professional advertisements.
- Develop horse profiles.
- Draft follow-up communications.
- Generate buyer question lists.
- Organise photographs, video
- Write realistic suitability descriptions

Improve website

- Plan content calendars.
- Write educational articles.
- Set up keywords and taglines
- Set up online forms

Marketing

- Build client resources.
- Build email sequences and automation systems.
- Create and place QR codes on promotional material.
- Create article ideas.
- Create brochures.
- Develop lead magnets, sales funnels.
- Develop newsletter topics.
- Draft social media content.
- Generate promotional ideas.

Riding schools

- Build rider progress templates.
- Build safety checklists.
- Create beginner rider guides.
- Create educational handouts and newsletter.
- Create stable etiquette guides.
- Develop lesson outlines.
- Develop school holiday programs.
- Draft clinic information.
- Prepare student on-boarding material.
- Write horse care resources.

If you read this list carefully, you may notice something. Very few of these examples are really about artificial intelligence. Most are about communication.

Rather like horses, really :)

Glossary

Action plan: A structured set of steps designed to move from a goal to a practical outcome. In this book, the personal AI action plan helps horse people identify one real problem and take one clear first step toward solving it.

Agent: A software system designed to perform tasks on behalf of a user.

AI Agent: An AI system that can carry out a sequence of actions to achieve a goal.

Algorithm: A set of rules used by software to decide what happens next.

Artificial Intelligence (AI): Computer systems that assist with writing, planning, organising information, answering questions, and generating content.

Aspect ratio: The shape of an image, such as square, portrait, or landscape.

Atticus: Book formatting software used to prepare print and ebook editions.

Automation: A process that happens automatically once it has been set up.

Avatar: A visual representation of a person or brand online.

Blog: A regularly updated section of a website containing articles or information.

Book Brush: A design tool often used for book graphics, mock-ups, and promotional images.

Booking system: Software that allows people to schedule appointments, lessons, events, or consultations.

Brand: The overall identity and reputation of a person, organisation, or business.

Buyer persona: A description of an ideal customer used to guide communication and marketing.

Call to action (CTA): A request encouraging readers to take a specific action.

Canva: A design platform used to create graphics, worksheets, brochures, and marketing materials.

Caption: Text accompanying a social media image or post. Often the primary communication vehicle on platforms like Facebook and Instagram. AI can help draft captions quickly from a brief description of the image or topic.

Chatbot: A software programme designed to hold a text-based conversation with a user. ChatGPT and Claude are advanced examples, but simpler chatbots are also found on many websites handling basic customer enquiries.

ChatGPT: An AI assistant that helps with writing, planning, brainstorming, and research.

Cinematic: An image style designed to resemble a scene from a film.

Claude: An AI assistant often used for writing and editing.

Clipchamp: A free video editing tool, built into Windows 11 and available as a web app for Mac users with a Microsoft account. Includes templates for social media formats, automatic captions, text to speech, and a music library, with no watermark on exported videos.

Close up: An image focused tightly on a subject.

Cloud storage: Online storage for files, photographs, and documents.

Compress / Compression: The process of reducing a file's size to make it easier to store, email, or upload. Image compression reduces file size by simplifying or removing some image data. Some compression is barely noticeable; too much results in a visibly lower quality image.

Content: Information created for education, communication, or marketing.

Content calendar: A schedule showing when content will be published.

Context: Background information supplied to improve AI responses.

Conversion: When a visitor takes a desired action such as joining a mailing list or making a purchase.

Copyright: The legal right of a creator to control how their original work is used, copied, or distributed. Using a photograph, image, or piece of writing found online without the explicit permission of the creator is a breach of copyright, regardless of whether it carries a watermark. Always use your own photographs or images licensed for commercial use.

Copywriting: Writing designed to persuade, educate, or encourage action.

Course: A structured educational programme delivered online or in person.

CRM: Customer Relationship Management software used to organise contacts and communication.

Crop: To trim the edges of an image to remove unwanted areas, improve composition, or resize it to fit a specific space. Cropping does not reduce the resolution of the remaining image but does reduce the total number of pixels available.

CSS: A language used to control the appearance of websites.

CSV file: Stands for comma-separated values. A simple file format used to store a list of contacts or data in rows and columns, similar to a basic spreadsheet. Each row represents one person or record, and each column holds a piece of information such as first name, last name, and email address. CSV files are commonly used to move contact lists between programs and platforms. The file name ends in .csv.

Depth of field: The amount of background blur in an image.

Digital asset: Anything valuable that exists in digital form.

Digital download: A file that can be downloaded from the internet.

Digital presence: How a person or business appears online.

Domain name: The web address used to locate a website.

Download: To transfer a file from the internet or another device onto your own device. Saving a photo, receiving an email attachment, or installing an app are all forms of downloading. The opposite is uploading, which sends a file from your device to another location.

DPI: Stands for dots per inch. Describes how densely pixels are packed when an image is printed. 72 DPI is standard for screens and social media. 300 DPI is the minimum required for professional print quality.

Draft: An unfinished version of a document.

Ebook: A book distributed in digital format.

Edit: To review and improve existing written content by correcting errors, refining wording, improving clarity, or adjusting tone and structure. In the context of AI, editing is one of the most practical and reliable uses of the technology. You can ask AI to edit your own writing for grammar, flow, and readability, or to edit AI generated content so that it sounds more like you before it is published or sent.

Email list: A collection of email subscribers who have agreed to receive communication.

Email sequence: A planned series of emails sent automatically.

Embed: To place or insert something directly into another file, document, or platform so that it becomes a permanent part of that item rather than existing as a separate file or external link. A common example is a YouTube video embedded into a website, where the video plays directly on that page rather than requiring the viewer to visit YouTube separately.

EPUB: A common ebook file format.

Evergreen content: Content that remains useful for a long period of time.

Export: The process of saving a finished file from a software programme in a format suitable for sharing or publishing. For example, exporting a video from InVideo AI as an MP4 file, or exporting a design from Canva as a PDF.

FAQ: Frequently Asked Questions.

File format: The type of file a document is saved as, indicated by the extension at the end of the file name. Common formats include PDF for finished documents intended for sharing, DOCX for editable Word documents, and EPUB for ebooks. The same content can often be saved in multiple formats depending on its intended use.

File size: The amount of storage space a file takes up on a device or online. A large high resolution image takes up more file size than a small low resolution one. File size is measured in kilobytes, megabytes, or gigabytes.

Fiverr: A freelance marketplace for hiring specialists.

Format: The structure, layout, or presentation of content. In design, format refers to how information is arranged on a page or screen. In file management, format describes the type of file a document is saved as (such as PDF, DOCX, or PNG).

Funnel: The process through which a potential client moves from first contact to becoming a customer.

GEO: Generative engine optimization: writing so AI tools like ChatGPT and Google's AI Overviews find and quote your work when readers seek guidance

Generative AI: AI capable of creating text, images, audio, video, or code.

Google Calendar: An online calendar used for scheduling and reminders.

Gumroad: A platform for selling digital products.

Hallucination: When AI presents incorrect information as though it were true. In practice, AI may confidently name a sire that never existed, or describe a breed trait with complete authority that is completely wrong. Always verify facts that matter.

Hero: A large, prominent image or visual element placed at the top of a webpage, social media post, or promotional material, designed to immediately capture attention and communicate the main message or brand identity.

High resolution: An image suitable for large displays or printing.

Homepage: The main page of a website.

Hook: The opening line or image of a video, designed to stop someone scrolling and make them want to keep watching. The first few seconds of a video matter more than the rest combined.

HorseRecords: A recommended horse management and record-keeping system.

HTML: A language used to structure web pages.

Hyperlink: A clickable word, phrase, or image that takes the reader directly to another webpage, document, or file. Often simply called a link.

Image generation: Creating images using AI from written instructions.

Image prompt: The written description used to generate an AI image.

Image recognition: The ability of AI to identify and interpret the contents of a photograph.

Integration: Connecting systems so they work together.

Iteration: Repeatedly improving something through multiple versions.

Intro: A short opening segment at the start of a video or podcast episode, often used to introduce the creator, the topic, or the brand. May include a brief animation, music, or spoken introduction.

InVideo AI: An online video creation tool that generates a draft video from a written prompt, including suggested footage, narration, and music. Free use is limited, with additional features available through paid credits.

JavaScript: A language used to add functionality to websites.

JPEG / JPG: The most common image file format. JPEGs are compressed to keep file sizes manageable, making them ideal for websites, email, and social media. Some image quality is lost in the compression process, so JPEGs are not always the best choice for professional print work.

JSON: A structured format used for sharing data.

KDP: Amazon Kindle Direct Publishing.

Keyword: A word or phrase used to help people find information.

Landing page: A webpage designed around a single goal.

Large Language Model (LLM): A type of AI trained to understand and generate language.

Lead: A person who may become a future client or customer.

Lead magnet: A valuable free resource offered in exchange for an email address or contact details. Examples include buyer guides, checklists, educational ebooks, horse care resources, and planning worksheets.

Lighting: Instructions controlling how bright, dark, or dramatic an image appears.

Lightroom: Adobe's photo management and editing tool, designed primarily for adjusting colour, exposure, and tone across large collections of images.

llms.txt: A plain text file placed on a website to help AI search tools and language models quickly understand what the site is about and where its most important content can be found.

Manuscript: The working version of a book before publication.

Markdown: A simple formatting language used in many online systems.

Megapixels: A measure of how many pixels a camera captures in a single photograph. One megapixel equals one million pixels. The higher the megapixel count, the more detail the image contains and the larger it can be printed without losing quality.

Metadata: Information that describes other information, such as author, date, keywords, or file details.

Midjourney: An AI image generation platform.

Mockup: A visual representation showing how a product may look when finished.

Narration: Spoken or written commentary that guides a viewer through a video or presentation. In AI video tools, narration is generated from your script and read aloud by a synthetic voice, or used as on-screen text.

Navigation menu: Links that help visitors move around a website.

Newsletter: Information sent directly to subscribers by email.

Niche: A specialised area of interest, expertise, or business.

Online calendar: A calendar stored and managed online.

Opt in: Giving permission to receive communication such as newsletters or updates.

Output: The result produced by AI.

Outro: A short closing segment at the end of a video or podcast episode, often used to thank the audience, prompt them to subscribe or follow, or direct them to more information. May include a brief animation, music, or spoken sign-off.

PDF: Portable Document Format, a commonly used document file type.

Perplexity: An AI research assistant.

Personal brand: The reputation and perception associated with an individual.

Photo-realistic: An image style designed to look like a real photograph.

Photoshop: Adobe's professional image editing software, widely used for retouching and manipulating photographs.

Pixelated: An image is described as pixelated when the individual pixels that make up the photograph become visible to the naked eye, causing the image to appear blocky, blurry, or broken up rather than smooth and clear. This most commonly happens when a small or low resolution image is enlarged beyond its original size. A pixelated image is unsuitable for professional print use.

Pixels: The tiny individual dots of colour that together make up a digital image. The more pixels an image contains, the more detail and clarity it holds.

Placeholder: A clearly marked position in a document or design indicating where specific content needs to be added before publication.

Platform: A service used to publish, communicate, store, or sell information.

PNG: A common image file format that preserves more detail than a JPEG and supports transparent backgrounds. PNGs are larger in file size but better suited to logos, graphics, and images where background transparency is needed.

Podcast: Audio content distributed online.

Polish: A prompt instruction asking AI to improve readability and presentation.

Portrait: An image focused on a person or animal.

Prompt: The instruction or question given to an AI system.

Prompt block: A reusable prompt provided as an example.

Prompt chain: A sequence of prompts used to refine results. Think of it as building a conversation. You start broad, then narrow down, the same way you might work through a training problem with a horse rather than asking for everything at once.

QR code: A small square pattern of black and white dots that links a printed or digital item directly to a website, product page, or online resource. Scanned using a smartphone camera, a QR code removes the need for someone to type a web address manually.

Reel: A short, vertical video, usually under 60 seconds, designed for platforms such as Facebook, Instagram, and TikTok. Reels are shown to a wider audience than standard posts, which makes them useful for visibility.

Render: The final image or output produced after processing your instructions or design choices.

Repurposing: Using the same information in multiple formats. One well-written stallion profile can become a Facebook post, a one-page flyer, a website bio, and a mare owner information pack. AI handles the conversion. You handle the quality check.

Research assistant: An AI used to gather, organise, and explain information.

Resolution: A measure of how much detail an image contains. Higher resolution means a sharper, clearer image suitable for print. Lower resolution is fine for screens but not for professional printing.

Rewrite: Creating a new version while preserving meaning.

Role prompt: A prompt that asks AI to adopt a specific role.

Screen: A prompt instruction asking AI to review and identify issues or opportunities.

Screenshot: A photograph of whatever is currently displayed on your screen, saved as an image file. Useful for sharing information, capturing records, or showing someone exactly what you are seeing on your device. On most devices, screenshots are taken using a simple button combination.

Search engine: A system used to find information online.

SEO: Search Engine Optimisation.

Sharp focus: An image instruction requesting crisp detail.

Shot list: A simple plan listing the specific photos or video clips needed for a video, in the order they will appear. Created before filming so that nothing important is missed.

Show notes: A written summary that accompanies a podcast episode, usually including a description of the topic, key points discussed, and any links mentioned. Helps new listeners decide whether to play an episode and assists with search visibility.

Social media: Platforms used to share content and communicate online.

Spam filter: An automatic system that identifies and diverts unwanted or suspicious emails away from your inbox. Emails generated by AI and sent in bulk can sometimes be incorrectly flagged by spam filters and never reach the recipient.

Spotify for Podcasters: A platform for publishing podcasts.

Subscriber: A person who has chosen to receive ongoing communication.

Subscription: A regular payment — usually monthly or annual — that gives access to a software platform or service. Most AI tools including ChatGPT and

Claude offer a free version with limitations and a paid subscription with greater capability.

Substack: A publishing platform for newsletters and articles, combining email newsletters with the ability to publish audio and video content directly to subscribers. Subscribers receive new content by email, rather than relying on a social media algorithm to show it to them.

Sweep: A prompt instruction asking AI to review an entire document or project.

Template: A reusable structure for documents, forms, or content. The horse industry equivalent: a standard foal registration form, a client welcome pack, or a mare owner questionnaire. Once it is built, you never have to start from scratch again.

Text overlay: Words placed on top of a video, used to convey information to viewers who are watching without sound. Many people watch Reels with the sound off, so overlays often carry the main message.

Text to speech: A feature that converts written text into a spoken voiceover automatically, using a computer-generated voice. Useful for adding narration to a video without recording your own voice.

Token: A unit of text processed by AI.

Tool directory: A reference list of useful software and services.

Training data: The large collection of existing text, images, and information that an AI system learned from during its development. Because AI training data has a cutoff date and may contain limited horse-specific material, AI can sometimes produce inaccurate information on specialist equine topics. Always verify facts that matter.

Two-factor authentication: A security process that requires two separate steps to log into an account — typically your password plus a code sent to your phone or email. Adds an important extra layer of protection to your accounts.

Ultra detailed: An image instruction requesting a high level of detail.

Unzipping a zip file: The process of extracting the contents of a zip file so that the individual files inside become accessible and usable again. On Windows, right-click the zip file and select Extract All. On a Mac, double-click and the contents appear automatically in a new folder. On a smartphone, a file manager app is usually required.

Upload: To transfer a file from your own device to the internet, a platform, or another device. Sharing a photo on social media, sending an email attachment, or adding an image to a website are all forms of uploading. The opposite is downloading.

Upscaling: The process of increasing the size of an image using software. Standard upscaling just stretches existing pixels, reducing quality. AI upscaling tools such as Topaz Gigapixel do a better job by intelligently filling in detail, but cannot fully replace a high resolution original.

URL: A web address leading to a webpage or resource.

Upscaling tools: Software that increases the size of an image using artificial intelligence to add detail rather than stretching pixels.

Vertical format: An image or video orientation that is taller than it is wide, using a 9:16 aspect ratio. Required for Facebook Reels, Instagram Reels, and TikTok. Portrait-orientation photos fit this format without cropping.

Video script: A written outline or script for a video.

Voice to text: Technology that converts spoken words into typed text. Increasingly useful for horse people who want to dictate notes, messages, or records hands-free while working in the stable or paddock. Available on most smartphones and within many AI platforms.

Watermark: A visible logo or text overlay applied to an image or video, usually indicating the creator or originating platform. Free versions of AI video tools such as InVideo AI often add a watermark to exported files. Paid plans typically remove it.

Webinar: A seminar or educational presentation delivered online.

Website builder: Software used to create websites without extensive coding.

White paper: A detailed report explaining a topic, idea, or proposal.

Wide angle: An image style showing a broader field of view.

Wording overlay: Text placed directly onto a promotional image — such as a stud fee, event date, location, or contact details. AI can help generate the wording; design tools such as Canva allow you to place it onto your photographs.

WordPress: A popular website publishing platform.

Workflow: The step-by-step process used to complete a task.

Worksheet: A structured document designed to guide learning or action.

XML: A structured data language used by some software systems.

YouTube: A video platform used for education, marketing, and communication.

Zip file: A single compressed file that contains one or more other files or folders packaged together to make them smaller and easier to store or share. Zip files are identified by the .zip extension at the end of the file name. Think of it as a suitcase — everything packed neatly in one container, taking up less space than if each item were carried separately.in one attachment.

Also by Jeanette Gower

Practical wisdom for all horse lovers

Homemade Dog Food Recipes

Feeding a dog well is one of the simplest and most meaningful ways owners show care, yet it can be difficult to know where to begin. In *Homemade Dog Food Recipes*, Jeanette Gower offers practical, balanced guidance designed for real homes and everyday routines.

It is more than just a recipe book. With clear instructions and economical ingredients, it helps owners build confidence in preparing meals that are science based and support long term well-being across different life stages.

The Rannock Legacy

In 2023, Rannock's induction into the Australian Stock Horse Society Hall of Fame marked the completion of a journey that began generations earlier. In *The Rannock Legacy*, Jeanette Gower records how one stallion's influence travelled across time, shaping horses, families, and breeding decisions far beyond his own years.

Drawing on careful research, photographs, and the recollections of those who knew his descendants, the book preserves a living thread within the breed. It is a record of continuity, respect, and the quiet transmission of qualities that still stand in paddocks and pedigrees today.

Buy The Right Horse

Buying a horse is exciting and full of promise, yet it is also a decision that shapes the years ahead. In *Buy The Right Horse*, Jeanette Gower offers a clear framework to help riders assess honestly and choose with confidence.

Drawing on decades of experience, she explains why suitability extends far beyond appearance or potential. Temperament, training, management, rider expectations, and future plans must align for a partnership to stay safe and satisfying.

Through structured questions, practical checklists, and real world examples, readers learn to recognise warning signs, communicate clearly with sellers, and interpret situations with confidence, grounding emotion in understanding.

Sell Your Horse The Right Way

A natural companion to *Buy The Right Horse*, this book helps buyers take responsibility for creating lasting matches that serve the rider, the seller, and most importantly, the horse. Jeanette addresses the realities sellers face, including emotional attachment, financial pressure, difficult conversations, and the temptation to overlook warning signs.

This book helps you set a fair, evidence-based price, write accurate advertisements that attract suitable buyers, and navigate private sales and auctions with clarity. You will also find practical checklists for enquiries, viewings, trials, and vet checks, along with guidance on recognising red flags early and avoiding preventable problems.

The result is steady guidance that helps sellers protect their reputation, act responsibly, and create matches that hold together.

The Thinking Horse Breeder

Breeding horses asks more than enthusiasm. It asks for judgement, patience, and decisions that take years to prove themselves.

In *The Thinking Horse Breeder*, Jeanette Gower distills more than five decades of practical experience into a guide that helps breeders understand why some programs endure while others falter. Covering selection of stock, planning matings, inheritance, conformation, temperament, and the raising of young horses, the book moves step by step through the realities that shape long term success. Complex ideas are made usable, always grounded in what happens in paddocks, not just on paper.

With over 150 colour photographs, readers return to this work again and again as their knowledge deepens. What first appears straightforward reveals further layers of meaning with experience, making it a trusted companion from beginner stages through to established breeding operations.

Also available in Spanish as *Cría Equina Inteligente*, and as an *audiobook* edition, this modern classic continues to guide breeders with the same careful guidance on breeding decisions, planning, and long term responsibility that has supported owners around the world.

All books available on Amazon and leading retailers.

Can I please ask you a favour?

Has this book given you even one useful idea, saved you an hour, or simply made you feel more confident in managing the complexity of horses and everything that comes with them?

I would be genuinely grateful if you would take a few minutes to leave a review on Amazon or Goodreads. Reviews matter enormously for independent authors. They help other horse people find the book, and they help me keep writing.

You do not need to write much. A sentence or two about what was useful, what surprised you, or how you plan to use what you have learned is more than enough.

About the author

Jeanette Gower's decades of experience span muddy paddocks, long nights foaling, and enough saddle time to write a book (literally). Based on the practicalities of horse care and breeding, Jeanette has made it her mission to help people make better decisions around horse ownership, from buying the right horse to building a lifelong partnership based on trust, ethics, and a lifetime of being around them.

Jeanette's approach is grounded in responsible horse ownership. *Buy the Right Horse* is her way of cutting through the fluff and guiding readers toward thoughtful, informed decisions. It's not about chasing ribbons or quick fixes; it's about making choices that prioritize the welfare of the horse and build stronger partnerships.

Jeanette is a voice of reason in an industry often clouded by trends and misconceptions. She advocates for transparency in breeding practices, ethical horse sales, and ongoing education for owners, because, as she says, "the learning never stops."

Her writing style blends hard-earned wisdom with an approachable tone, making complex ideas digestible for readers of all levels; expect practical advice delivered with a few home truths.

Jeanette's philosophy? "Think before you buy. Think harder before you breed."

She believes in building confidence through knowledge. Her work encourages readers to embrace both the highs and the challenges of horse ownership with eyes wide open. She's not afraid to call out poor practices, but she does so with the heart of a mentor, offering actionable solutions.

Her deep commitment to horse welfare shines through in every anecdote, every lesson, and every strong opinion she shares, because when it comes to horses, there's no room for half measures.

So whether you're a first-time buyer or a seasoned breeder looking for a fresh perspective, Jeanette invites you to join her journey toward more mindful, meaningful horse ownership. After all, as she's fond of saying, "Every horse tells a story - the horse is a reflection of *you*."

***

In 2026 Jeanette built *The Thinking Horseman Series*, and added this book to the collection. Says Jeanette:

"I created *The Thinking Horseman Series* because I saw a growing gap between traditional horsemanship and the needs of today's horse owners. For generations, knowledge passed naturally from experienced horsemen to the next, but today, many modern owners come to horses through recreation rather than a lifetime in the industry.

My goal is to bridge that gap: preserving the practical wisdom, ethics, and common sense of traditional horsemanship while combining it with modern knowledge, so better informed people can ask better questions, make stronger decisions, build deeper partnerships, and ultimately create better lives for the horses we all love."

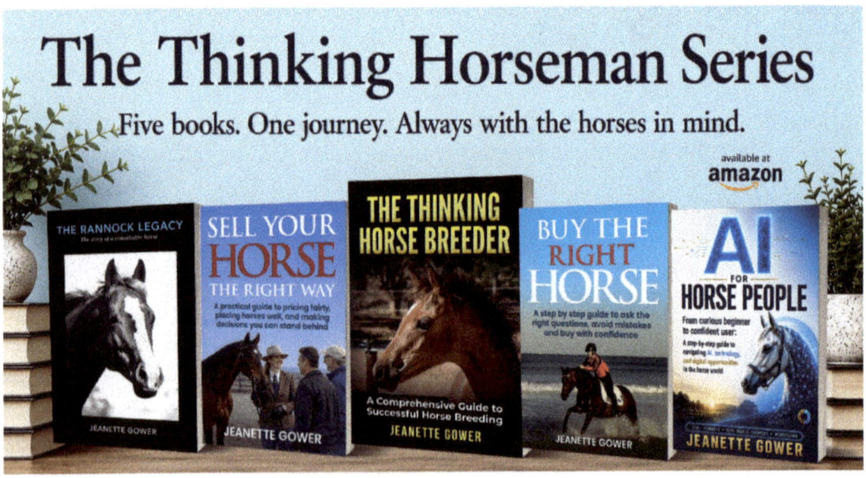

You can contact the author at jeanette.gower@gmail.com or on Facebook at Chalani Australian Stock Horses,
https://www.facebook.com/ChalaniStud/

www.ingramcontent.com/pod-product-compliance
Lightning Source LLC
LaVergne TN
LVHW050912210926
848482LV00012B/415